Compact Textbooks in Mathematics

This textbook series presents concise introductions to current topics in mathematics and mainly addresses advanced undergraduates and master students. The concept is to offer small books covering subject matter equivalent to 2- or 3-hour lectures or seminars which are also suitable for self-study. The books provide students and teachers with new perspectives and novel approaches. They may feature examples and exercises to illustrate key concepts and applications of the theoretical contents. The series also includes textbooks specifically speaking to the needs of students from other disciplines such as physics, computer science, engineering, life sciences, finance.

- **compact:** small books presenting the relevant knowledge
- **learning made easy:** examples and exercises illustrate the application of the contents
- **useful for lecturers:** each title can serve as basis and guideline for a semester course/lecture/seminar of 2-3 hours per week.

Ulrich Pinkall • Oliver Gross

Differential Geometry

From Elastic Curves to Willmore Surfaces

 Birkhäuser

Ulrich Pinkall
Department of Mathematics
Technical University of Berlin
Berlin, Germany

Oliver Gross
Department of Mathematics
Technical University of Berlin
Berlin, Germany

ISSN 2296-4568   ISSN 2296-455X (electronic)
Compact Textbooks in Mathematics
ISBN 978-3-031-39837-7   ISBN 978-3-031-39838-4 (eBook)
https://doi.org/10.1007/978-3-031-39838-4

This work was supported by Deutsche Forschungsgemeinschaft and Technische Universität Berlin.

This book is published under the imprint Birkhäuser, www.birkhauser-science.com by the registered company Springer Nature Switzerland AG
The registered company address is: Gewerbestrasse 11, 6330 Cham, Switzerland

Paper in this product is recyclable.

Preface

This book is based on a course on the differential geometry of curves and surfaces at Technische Universität Berlin in the spring term of 2020. The 13 chapters roughly reflect the 13 weeks of that term.

The pioneers of differential calculus like Newton, Bernoulli, and Euler immediately applied their ideas to questions about curves and surfaces. In 1673, Newton defined the curvature κ of a plane curve, and in 1691, Jacob Bernoulli characterized elastic plane curves (cf. [30], [25], or [3] for a historical overview), i.e. curves that minimize the bending energy $\int \kappa^2 \, ds$ among all curves held fixed at their end points. In 1859, Kirchhoff showed that the tangent vector of an elastic curve follows the motion of the axis of a spinning top [19]. Even today, many applications of differential geometry of curves in other sciences (ranging from the coiling of DNA strands (cf. [37]) to the modeling of hair for computer-generated imagery (cf. [4])) are centered around elastic curves. Our approach to curve theory emphasizes its connections to the calculus of variations and we will explore elastic curves quite thoroughly.

There is also a dynamic aspect of curve theory, where deformations of curves in time are studied. In 1906, Da Rios, a student of Levi-Civita, derived an evolution equation [33], the so-called filament flow, for space curves that models the motion of vortex filaments in a fluid (Sect. 5.3). In 1932, Levi-Civita wrote the equations satisfied by filaments that do not change shape under this flow [24]. In 1991, Langer and Perline showed that the possible shapes of such filaments are given by elastic curves [31], a fact that had escaped Levi Civita. Already in 1972, Hasimoto had shown that the filament flow is a so-called Soliton equation [15]. Even today this insight remains a source of ongoing inspiration for curve theory (see [10] for a survey).

While minimizing the length of a curve results in straight line segments, minimizing the area $\int \det$ of a surface with a given boundary curve leads to a rich class of surfaces, the so-called minimal surfaces (Sect. 12.4). Already in 1744, Euler proved that the catenoid minimizes area among all surfaces of revolution with prescribed boundary circles (cf. [13]). Minimizing area while fixing the enclosed volume leads to surfaces with constant mean curvature H (Sect. 12.5). For a surface, the analog of the bending energy $\int \kappa^2 \, ds$ of a curve is the so-called Willmore functional $\int H^2 \det$ (Sect. 13.1). In the context of surfaces, the analog of an elastic curve is a so-called Willmore surface, whose equation we derive in Sect. 13.2. It was a major milestone in differential geometry when in 2012 Marques and Neves proved the so-called Willmore conjecture (cf. [26]), which states that for any torus in $\mathbb{R}^3$ the Willmore functional has to be at least $2\pi^2$.

A pervasive theme in differential geometry is the interplay between curvature and topology. In Sect. 3.4, we will show that the integral $\int \kappa \, ds$ of the curvature of a closed plane curve γ equals 2π times an integer, the so-called tangent winding number of γ. In Sect. 3.6, we follow Whitney and Graustein who proved in 1937 that this integer characterizes the connected components of the space of all closed plane curves [45]. In the context of surfaces, the analog of this result is the Gauss-Bonnet theorem for closed surfaces [8], which we prove in Sect. 10.2.

The only prerequisites for this book are the calculus of several variables including the transformation formula for integrals, the Picard-Lindelöf theorem for ordinary differential equations, and Green's theorem from vector calculus. Neither manifolds nor results from functional analysis are needed. Variational problems under constraints are accessible with these prerequisites because our definition of a critical point under constraints (Definition 2.19) is slightly stronger than the usual one. Similarly, our ability to discuss the genus of closed surfaces without diving into algebraic topology can be traced back to our definition of a compact domain with smooth boundary in $\mathbb{R}^2$ (Definition 6.1). Our definition is intuitive and equivalent to the standard one, but proving this equivalence would need serious additional work.

Berlin, Germany   Ulrich Pinkall
January 2021

Acknowledgement

This work was funded by the Deutsche Forschungsgemeinschaft (DFG—German Research Foundation)—Project-ID 195170736—TRR109 "Discretization in Geometry and Dynamics" and the Open Access Publication Fund of TU Berlin. Additional support was provided by SideFX software. We sincerely thank Javier Villegas, Geoff Goss, and William Irvine for allowing us to use their images.

Contents

Part II Surfaces

Part I
Curves

Curves in $\mathbb{R}^n$

Differential Geometry studies smoothly curved shapes, called *manifolds*. One-dimensional shapes are called *curves* and two-dimensional shapes are called *surfaces*. In this chapter we look at curves in n-dimensional Euclidean space. The basic properties of curves in $\mathbb{R}^n$ (length, tangent, bending energy) were explored right after the invention of calculus by Newton, Bernoulli and Euler.

1.1 What is a Curve in $\mathbb{R}^n$?

Since many interesting curves (for example a figure eight) have self-intersections, it is not a good idea to define a curve as a special kind of subset in $\mathbb{R}^n$. Intuitively, a curve is something that can be traced out ("parametrized") as the path of a moving point (cf. Fig. 1.1).

Definition 1.1

A **curve** in $\mathbb{R}^n$ is a smooth map $\gamma \colon [a, b] \to \mathbb{R}^n$ such that its velocity vector $\gamma'(x)$ never vanishes, i.e.

$$\gamma'(x) \neq 0$$

for all $x \in [a, b]$.

▶ **Remark 1.2** If $M \subset \mathbb{R}^n$ is an arbitrary subset, then a map $f \colon M \to \mathbb{R}^k$ is called **smooth** (or C^∞) if there is an open set $U \subset \mathbb{R}^n$ with $M \subset U$ and an infinitely often differentiable map $\tilde{f} \colon U \to \mathbb{R}^k$ such that $f = \tilde{f}|_M$ (cf. Appendix A.1). Instead of a closed interval $[a, b]$ one could also allow an open or semi-open interval (or even a finite union of intervals) as the domain of definition for a curve. The only problem that would arise is that then the integral of a smooth function would not always be defined. For all of our applications we can stick to closed intervals.

© The Author(s) 2024
U. Pinkall, O. Gross, *Differential Geometry*, Compact Textbooks in Mathematics,
https://doi.org/10.1007/978-3-031-39838-4_1

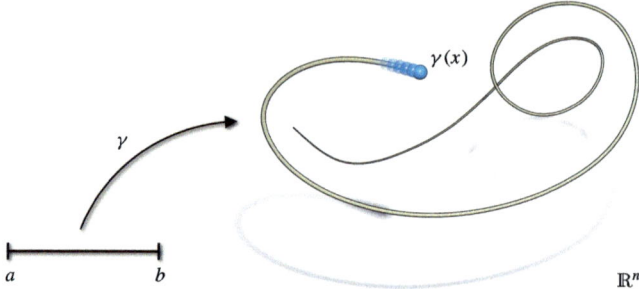

Fig. 1.1 A curve can be described as the trajectory of a particle moving in space. The particles position at time x is given by $\gamma(x)$

Definition 1.3

A curve $\gamma : [a, b] \to \mathbb{R}^n$ is called **closed** if γ can be extended to a smooth map $\tilde{\gamma} : \mathbb{R} \to \mathbb{R}^n$ with *period* $b - a$, which means

$$\tilde{\gamma}(x + (b - a)) = \tilde{\gamma}(x)$$

for all $x \in \mathbb{R}$.

Example 1.4

(i) The quarter circle is a curve:

$$\gamma : \left[-\tfrac{1}{\sqrt{2}}, \tfrac{1}{\sqrt{2}}\right] \to \mathbb{R}^2, \ \gamma(x) = \begin{pmatrix} t \\ \sqrt{1 - x^2} \end{pmatrix}.$$

(ii) Another version of the quarter circle is also a curve:

$$\gamma : \left[\tfrac{\pi}{4}, \tfrac{3\pi}{4}\right] \to \mathbb{R}^2, \ \gamma(x) = \begin{pmatrix} \cos x \\ \sin x \end{pmatrix}.$$

(iii) The full circle

$$\gamma : [0, 2\pi] \to \mathbb{R}^2, \ \gamma(x) = \begin{pmatrix} \cos x \\ \sin x \end{pmatrix}$$

is a closed curve with period 2π. It can be extended to

$$\tilde{\gamma} : \mathbb{R} \to \mathbb{R}^2, \ \gamma(x) = \begin{pmatrix} \cos x \\ \sin x \end{pmatrix}.$$

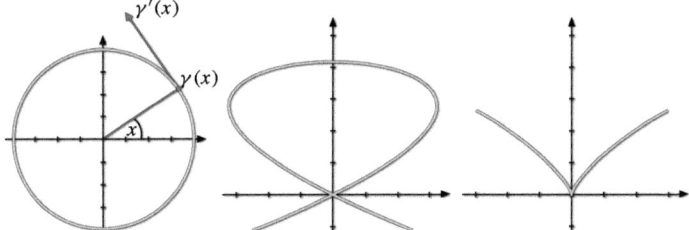

Fig. 1.2 A circle *(left)*, the Cartesian leaf *(middle)* and Neil's parabola *(right)*

(iv) The **Helix** is a curve:

$$\gamma : [a, b] \to \mathbb{R}^3, \ \gamma(x) = \begin{pmatrix} \cos x \\ \sin x \\ x \end{pmatrix}.$$

(v) The **Cartesian leaf** (see Fig. 1.2) is a curve:

$$\gamma : [a, b] \to \mathbb{R}^2, \ \gamma(t) = \begin{pmatrix} x^3 - 4x \\ x^2 - 4 \end{pmatrix}$$

so that

$$\gamma'(t) = \begin{pmatrix} 3x^2 - 4 \\ 2x \end{pmatrix}.$$

(vi) **Neil's parabola** (see Fig. 1.2) is given by

$$\gamma : [a, b] \to \mathbb{R}^2, \ \gamma(t) = \begin{pmatrix} x^3 \\ x^2 \end{pmatrix}.$$

It is not a curve if $0 \in [a, b]$, because at $t = 0$

$$\gamma'(0) = \begin{pmatrix} 0 \\ 0 \end{pmatrix}.$$

For the purposes of geometry, the speed with which we run through a curve does not really matter, nor does the particular time interval $[a, b]$ that we use for the parametrization. However, we will always assume that our curves are *oriented*, so we want to keep track of the direction in which we run through the curve. This means that we are only interested in properties of a curve that do not change under orientation-preserving reparametrization (see Fig. 1.3):

Fig. 1.3 A reparametrization
of a curve is given by a
strictly increasing function
with nowhere vanishing
derivative which maps $[c, d]$
onto $[a, b]$

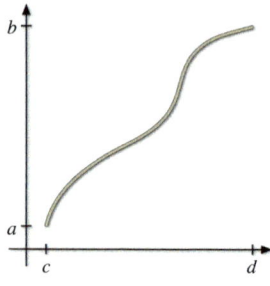

Definition 1.5

Let $\gamma : [a, b] \to \mathbb{R}^n$ and $\tilde{\gamma} : [c, d] \to \mathbb{R}^n$ be two curves. Then $\tilde{\gamma}$ is called an
orientation-preserving reparametrization of γ if there is a bijective smooth
map $\varphi : [c, d] \to [a, b]$ such that $\varphi'(x) > 0$ for all $x \in [c, d]$ and $\tilde{\gamma} = \gamma \circ \varphi$.

Example 1.6
For the two curves γ from Example 1.4 (i) and $\tilde{\gamma}$ from Example 1.4 (ii) we have $\tilde{\gamma} = \gamma \circ \varphi$ with

$$\varphi : \left[\tfrac{\pi}{4}, \tfrac{3\pi}{4} \right] \to \left[-\tfrac{1}{\sqrt{2}}, \tfrac{1}{\sqrt{2}} \right], \quad \varphi(x) = \cos x.$$

▶ **Remark 1.7** Orientation-preserving reparametrization is an equivalence relation
on the set of curves in $\mathbb{R}^n$. Although we are ultimately only interested in properties
shared by all curves in the same equivalence class, we will always work with a
particular representative curve γ.

1.2   Length and Arclength

The most simple numerical quantity that can be assigned to a curve as a whole is its
length.

Definition 1.8

Let $\gamma : [a, b] \to \mathbb{R}^n$ be a curve. Then the function

$$v : [a, b] \to \mathbb{R}, \ t \mapsto |\gamma'(t)|$$

is called the **speed** of γ and

$$\mathcal{L}(\gamma) := \int_a^b v$$

is called the **length** of γ.

The length of a curve does not change under reparametrization:

> **Theorem 1.9**
> *Suppose* $\gamma\colon [a,b] \to \mathbb{R}^n$ *and* $\tilde{\gamma}\colon [c,d] \to \mathbb{R}^n$ *are two curves such that* $\tilde{\gamma} = \gamma \circ \varphi$ *for some diffeomorphism* $\varphi\colon [c,d] \to [a,b]$. *Then* γ *and* $\tilde{\gamma}$ *have the same length.*

Proof. By the substitution rule, we have

$$\mathcal{L}(\tilde{\gamma}) = \int_c^d |(\gamma \circ \varphi)'| = \int_c^d |\gamma' \circ \varphi| \varphi' = \int_a^b |\gamma'| = \mathcal{L}(\gamma).$$

$\square$

Example 1.10

(i) For the half circle $\gamma\colon [0, \pi] \to \mathbb{R}^2$,

$$\gamma(x) = \begin{pmatrix} \cos x \\ \sin x \end{pmatrix}$$

we have $|\gamma'| = 1$ and therefore $\mathcal{L}(\gamma) = \pi$.

(ii) The line segment $\gamma\colon [a,b] \to \mathbb{R}^2$,

$$\gamma(x) = \begin{pmatrix} x \\ 0 \end{pmatrix}$$

has length $\mathcal{L}(\gamma) = b - a$.

Definition 1.11

A **rigid motion** of $\mathbb{R}^n$ is a map $g\colon \mathbb{R}^n \to \mathbb{R}^n$ of the form

$$g(\mathbf{y}) = A\mathbf{y} + \mathbf{b}$$

where $A \in O(n)$ is an orthogonal matrix and $\mathbf{b} \in \mathbb{R}^n$ is a vector.

Rigid motions are those transformations of the ambient space $\mathbb{R}^n$ which preserve distances between points. Two shapes that differ only by a rigid motion are said to be **congruent**. Matching the physical intuition for curves as trajectories of a particle moving in space, the length of a curve is invariant under rigid motions:

Theorem 1.12

Let $\gamma\colon [a, b] \to \mathbb{R}^n$ be a curve and $g\colon \mathbb{R}^n \to \mathbb{R}^n$ a rigid motion. Then

$$\mathcal{L}(g \circ \gamma) = \mathcal{L}(\gamma).$$

Proof. For $\tilde{\gamma} = g \circ \gamma$ we have $\tilde{\gamma} = A\gamma + \mathbf{b}$ and $\tilde{\gamma}' = A\gamma'$. Therefore,

$$\mathcal{L}(\tilde{\gamma}) = \int_a^b |A\gamma'| = \int_a^b |\gamma'| = \mathcal{L}(\gamma).$$

$\square$

Definition 1.13

Let $\gamma\colon [a, b] \to \mathbb{R}^n$ be a curve. Then the function

$$s\colon [a, b] \to \mathbb{R}, \ s(t) := \mathcal{L}\left(\gamma|_{[a,t]}\right) = \int_a^t |\gamma'|$$

is called the **arclength** function (or **arclength coordinate**) of γ.

In most situations however, the arclength function s itself is less useful than its derivative, the speed $s' = v = |\gamma'|$. Using only v, not s, we can define the derivative with respect to arclength:

Definition 1.14

Let $\gamma\colon [a, b] \to \mathbb{R}^n$ be a curve and $v = |\gamma'|$ its speed. Let $g\colon [a, b] \to \mathbb{R}^k$ be a smooth function. Then we define the **derivative with respect to arclength** of g as the function

$$\frac{dg}{ds} := \frac{g'}{v}$$

and the **integral over arclength** of g as

$$\int_a^b g \, ds := \int_a^b g \, v.$$

▶ **Remark 1.15** Once we have learned about 1-forms in Sect. 7.2, we will be able to interpret ds as a 1-form on $[a, b]$ and $\frac{dg}{ds}$ as quotient of 1-forms, just as it had been the dream of Leibniz. For now, they are just $\mathbb{R}^k$-valued functions on $[a, b]$.

Theorem 1.16

The arclength fucnction $s: [a, b] \rightarrow \mathbb{R}$ of a curve $\gamma: [a, b] \rightarrow \mathbb{R}^n$ is an orientation-preserving diffeomorphism of the interval $[a, b]$ onto the interval $[0, L]$ where $L = \mathcal{L}(\gamma)$. The reparametrization

$$\tilde{\gamma}: [0, L] \rightarrow \mathbb{R}^n, \ \tilde{\gamma} = \gamma \circ s^{-1}$$

has unit speed, i.e. $|\tilde{\gamma}'| = 1$.

▶ **Remark 1.17** It is common in the literature on curves to routinely assume that the curves under consideration have unit speed, usually expressed by saying that they are "**parametrized by arclength**". We will not do this here, for the following reasons:

(i) Making use of the derivative with respect to arclength defined in 1.14 gives us the same elegant formulas as they arise in the context of unit speed curves, without actually changing the parametrization.

(ii) When dealing with one-parameter families $t \mapsto \gamma_t$ of curves of varying length, one cannot assume that all curves γ_t are parametrized by unit speed. Therefore, in this situation one has to resort anyway to formulas that remain valid for arbitrary curves.

(iii) In the context of surfaces, there is no obvious analog for the unit speed parametrization of a curve. Therefore, habitual reliance on unit speed parametrizations makes the theory of surfaces look more different from the theory of curves than it actually is.

1.3   Unit Tangent and Bending Energy

Definition 1.18

For a curve $\gamma: [a, b] \rightarrow \mathbb{R}^n$, the normalized velocity vector field

$$T: [a, b] \rightarrow S^{n-1}, \ T = \frac{d\gamma}{ds} = \frac{\gamma'}{|\gamma'|}$$

is called the **unit tangent field** of γ.

Next to the length, the most important numerical quantity that can be assigned to a curve as a whole is its bending energy:

Definition 1.19

Let T be the unit tangent field of a curve $\gamma : [a, b] \to \mathbb{R}^n$. Then

$$\mathcal{B}(\gamma) = \frac{1}{2} \int_a^b \left\langle \frac{dT}{ds}, \frac{dT}{ds} \right\rangle ds$$

is called the **bending energy** of γ.

The bending energy is invariant under orientation-preserving reparametrization. The name comes from the following physical picture:

Consider a rod manufactured out of some elastic material in the shape of a thin cylinder of length L and radius ϵ. Then we bend the cylinder into the shape of a curve γ of length L. While doing this, we make sure that we do not force any twisting on the cylinder, for example we place the cylinder in a hollow tube with shape γ, leaving it free to untwist itself within the tube (see Fig. 1.4). Then, in the limit of $\epsilon \to 0$, the energy needed to bring the initially straight rod into its new shape will be proportional to $\mathcal{B}(\gamma)$.

In later sections we will find out what curves we obtain if we hold a curve fixed near its end points but otherwise let it minimize bending energy (cf. Fig. 2.3). We also will find a way to deal with twisting.

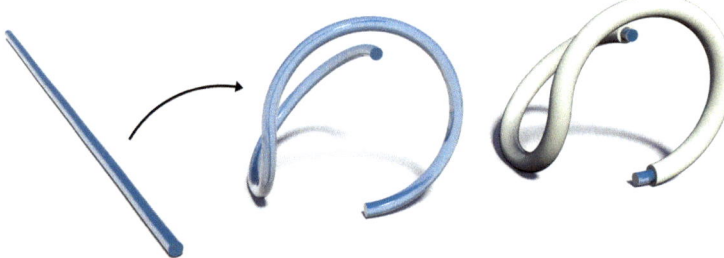

Fig. 1.4 A rod is bent into the shape of a curve. Then it is fixed in its position by a porcelain case within which it can untwist while staying in shape

Variations of Curves

2

Many important special curves γ arise by minimizing a certain variational energy $E(\gamma)$. For example, $E(\gamma)$ could be a linear combination of length and bending energy, in which case the curve is called an *elastic curve*. We are not only interested in minima but also in unstable energetic equilibria, possibly under constraints like fixing the curve near its end points. In this chapter we develop the basics of Variational Calculus. In particular, this allows us to explore elastic curves. Beyond straight lines and circles, these are the most important special curves in $\mathbb{R}^n$.

2.1 One-Parameter Families of Curves

On many occasions we will have to deal not only with individual curves $\gamma : [a, b] \to \mathbb{R}^n$ but with whole one-parameter families $t \mapsto \gamma_t$ of curves.

Definition 2.1

Let $g_t : [a, b] \to \mathbb{R}^k$ be a smooth map, defined for each $t \in [t_0, t_1]$. Then the **one-parameter family of maps** $[t_0, t_1] \ni t \mapsto g_t$ is called smooth if the map

$$[a, b] \times [t_0, t_1] \to \mathbb{R}^k, \ (x, t) \mapsto g_t(x)$$

is **smooth** (as always, in the sense of Remark 1.2).

Given a smooth one-parameter family

$$t \mapsto (g_t : [a, b] \to \mathbb{R}^k), \quad t \in [t_0, t_1]$$

of maps, also

$$t \mapsto g_t'$$

© The Author(s) 2024
U. Pinkall, O. Gross, *Differential Geometry*, Compact Textbooks in Mathematics,
https://doi.org/10.1007/978-3-031-39838-4_2

is a smooth one-parameter family of maps $g'_t \colon [a, b] \to \mathbb{R}^k$. The same holds for $t \mapsto \dot{g}_t$ where $\dot{g}_t \colon [a, b] \to \mathbb{R}^k$ is defined as

$$\dot{g}_t(x) := \frac{d}{d\tau}\bigg|_{\tau = t} g_\tau(x).$$

The dot and prime derivatives are just partial derivatives, so they commute by Schwarz's theorem:

Theorem 2.2

For a smooth one-parameter family of maps $t \mapsto g_t$, where $g_t \colon [a, b] \to \mathbb{R}^k$ we have

$$(g')^{\boldsymbol{\cdot}} = (\dot{g})'.$$

In our context, one-parameter families of maps will mainly arise as variations of a single map $g \colon [a, b] \to \mathbb{R}^k$:

Definition 2.3

A smooth one-parameter family $t \mapsto g_t$ of maps from M to $\mathbb{R}^k$ is called a **variation of a smooth map** $g \colon M \to \mathbb{R}^k$ if $t_0 < 0 < t_1$ and $g_0 = g$. In this context, we will also use the notation

$$\dot{g} := \dot{g}_0.$$

Our main interest is in variations of curves $\gamma \colon [a, b] \to \mathbb{R}^n$ (and the associated variations of derived quantities like the unit tangent or the length):

Definition 2.4

For a variation $t \mapsto \gamma_t$ of a curve $\gamma \colon [a, b] \to \mathbb{R}^n$ the map

$$Y := \dot{\gamma} \colon [a, b] \to \mathbb{R}^n$$

is called its **variational vector field**.

Suppose we have a smooth one-parameter-family $t \mapsto \gamma_t$ of curves $\gamma_t \colon [a, b] \to \mathbb{R}^n$, meaning that $\gamma'_t(x) \neq 0$ for all $x \in [a, b]$ and all $t \in [t_0, t_1]$. Then we can think of this family (just for the purpose of intuition, no need for further formal definitions) as a smooth map from $[t_0, t_1]$ into in the space $\mathcal{M}$ of all curves $\gamma \colon [a, b] \to \mathbb{R}^n$. The vector $\dot{\gamma}_t \in C^\infty([a, b], \mathbb{R}^n)$ can then be thought of as the "velocity vector" of that map at time t (see Fig. 2.1).

Fig. 2.1 A variation of a curve γ can be interpreted as a map into the space $\mathcal{M}$ of all curves $\gamma : [a, b] \to \mathbb{R}^n$

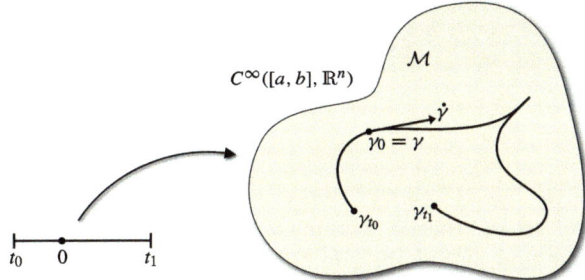

▶ **Remark 2.5** Throughout this whole book we will treat $C^\infty\left([a, b], \mathbb{R}^k\right)$ (and its analog in the context of surfaces) only as a vector space, based on notions from Linear Algebra. So, for example, we will indeed use the Euclidean inner product

$$\langle\langle g, h \rangle\rangle := \int_a^b \langle g, h \rangle$$

but we will never put any topology on $C^\infty([a, b], \mathbb{R}^k)$. This means that you will get confused if you try to interpret what we say based on notions from Functional Analysis. These notions have important applications in Differential Geometry, but they are not used at all in this book.

2.2 Variation of Length and Bending Energy

Given a variation $t \mapsto \gamma_t$ of a curve $\gamma : [a, b] \to \mathbb{R}^n$, we want to determine $\frac{d}{dt}\big|_{t=0} \mathcal{L}(\gamma_t)$ and $\frac{d}{dt}\big|_{t=0} \mathcal{B}(\gamma_t)$. We first compute the time derivative of the integrands of these integrals:

Theorem 2.6

Let $t \mapsto \gamma_t$ be a variation with variational vector field $Y : [a, b] \to \mathbb{R}^n$ of a curve $\gamma : [a, b] \to \mathbb{R}^n$ with speed $v = ds$ and unit tangent field T. Then the variation of ds is given by

$$\dot{ds} = \left\langle \frac{dY}{ds}, T \right\rangle ds.$$

Proof. Differentiating the equation $v_t = |\gamma_t'|$ with respect to t at $t = 0$ we obtain

$$\dot{v} = \frac{\langle \dot{\gamma}', \gamma' \rangle}{v} = \left\langle \frac{Y'}{v}, \gamma' \right\rangle = \left\langle \frac{dY}{ds}, T \right\rangle ds.$$

$\square$

Before we proceed to compute the rate of change for the bending energy integrand, note that (unlike the situation for partial derivatives), for a one-parameter family $t \mapsto \gamma_t$ the derivative with respect to t does not commute with the derivative with respect to arclength:

Theorem 2.7

Let $t \mapsto \gamma_t$ be a variation with variational vector field $Y : [a, b] \to \mathbb{R}^n$ of a curve $\gamma : [a, b] \to \mathbb{R}^n$ with speed $v = ds$. Then for any one-parameter family $t \mapsto g_t$ of functions $g_t : [a, b] \to \mathbb{R}^k$ with $g_0 =: g$ we have

$$\left(\frac{dg}{ds} \right)^{\cdot} = \frac{d\dot{g}}{ds} - \left\langle \frac{dY}{ds}, T \right\rangle \frac{dg}{ds}.$$

Proof. By Theorem 2.6,

$$\left(\frac{dg}{ds} \right)^{\cdot} = \left(\frac{g'}{v} \right)^{\cdot} = \frac{(g')^{\cdot}}{v} - \frac{\langle \frac{dY}{ds}, T \rangle v}{v^2} g' = \frac{d\dot{g}}{ds} - \left\langle \frac{dY}{ds}, T \right\rangle \frac{dg}{ds}.$$

$\square$

Theorem 2.8

Given a variation $t \mapsto \gamma_t$ with variational vector field $Y : [a, b] \to \mathbb{R}^n$ of a curve $\gamma : [a, b] \to \mathbb{R}^n$ with speed $v = ds$, the corresponding variation of the bending energy density is

$$\left(\frac{1}{2} \left\langle \frac{dT}{ds}, \frac{dT}{ds} \right\rangle ds \right)^{\cdot} = \left(\left\langle \frac{d^2Y}{ds^2}, \frac{dT}{ds} \right\rangle - \frac{3}{2} \left\langle \frac{dY}{ds}, T \right\rangle \left\langle \frac{dT}{ds}, \frac{dT}{ds} \right\rangle \right) ds.$$

Proof. Applying Theorem 2.7 to $g = \gamma$ we obtain

$$\dot{T} = \frac{dY}{ds} - \left\langle \frac{dY}{ds}, T \right\rangle T.$$

Using this, Theorem 2.6, the fact that $\langle T, T \rangle = 1$ implies $\left\langle \frac{dT}{ds}, T \right\rangle = 0$ and Theorem 2.7 with $g = T$ we obtain

$$\left(\frac{1}{2} \left\langle \frac{dT}{ds}, \frac{dT}{ds} \right\rangle ds \right)^{\cdot} = \left\langle \left(\frac{dT}{ds} \right)^{\cdot}, \frac{dT}{ds} \right\rangle ds + \frac{1}{2} \left\langle \frac{dT}{ds}, \frac{dT}{ds} \right\rangle \left\langle \frac{dY}{ds}, T \right\rangle ds$$

$$= \left\langle \frac{d\dot{T}}{ds} - \left\langle \frac{dY}{ds}, T \right\rangle \frac{dT}{ds}, \frac{dT}{ds} \right\rangle ds + \frac{1}{2} \left\langle \frac{dT}{ds}, \frac{dT}{ds} \right\rangle \left\langle \frac{dY}{ds}, T \right\rangle ds$$

$$= \left\langle \frac{d\dot{T}}{ds}, \frac{dT}{ds} \right\rangle ds - \frac{1}{2} \left\langle \frac{dT}{ds}, \frac{dT}{ds} \right\rangle \left\langle \frac{dY}{ds}, T \right\rangle ds$$

$$= \left\langle \frac{d^2 Y}{ds^2} - \left\langle \frac{dY}{ds}, T \right\rangle \frac{dT}{ds}, \frac{dT}{ds} \right\rangle ds - \frac{1}{2} \left\langle \frac{dT}{ds}, \frac{dT}{ds} \right\rangle \left\langle \frac{dY}{ds}, T \right\rangle ds$$

$$= \left\langle \frac{d^2 Y}{ds^2}, \frac{dT}{ds} \right\rangle ds - \frac{3}{2} \left\langle \frac{dT}{ds}, \frac{dT}{ds} \right\rangle \left\langle \frac{dY}{ds}, T \right\rangle ds$$

□

The proof of the following theorem is based on applying integration by parts repeatedly.

Theorem 2.9

Given a variation $t \mapsto \gamma_t$ with variational vector field $Y : [a, b] \to \mathbb{R}^n$ of a curve $\gamma : [a, b] \to \mathbb{R}^n$, the corresponding variations of the length and bending energy are

$$\frac{d}{dt}\bigg|_{t=0} \mathcal{L}(\gamma_t) = \langle Y, T \rangle |_a^b - \int_a^b \left\langle Y, \frac{dT}{ds} \right\rangle ds$$

$$\frac{d}{dt}\bigg|_{t=0} \mathcal{B}(\gamma_t) = \left(\left\langle \frac{dY}{ds}, \frac{dT}{ds} \right\rangle - \left\langle Y, \frac{d^2 T}{ds^2} + \frac{3}{2} \left\langle \frac{dT}{ds}, \frac{dT}{ds} \right\rangle T \right\rangle \right) \bigg|_a^b$$

$$+ \int_a^b \left(\left\langle Y, \frac{d^3 T}{ds^3} + 3 \left\langle \frac{dT}{ds}, \frac{d^2 T}{ds^2} \right\rangle T + \frac{3}{2} \left\langle \frac{dT}{ds}, \frac{dT}{ds} \right\rangle \frac{dT}{ds} \right\rangle \right) ds.$$

Proof. By Theorem 2.8,

$$\frac{d}{dt}\Big|_{t=0}\mathcal{L}(\gamma_t) = \int_a^b \dot{s}\,ds$$

$$= \int_a^b \left\langle \frac{dY}{ds}, T \right\rangle ds$$

$$= \int_a^b \left(\frac{d}{ds}\langle Y, T\rangle - \left\langle Y, \frac{dT}{ds} \right\rangle \right) ds$$

$$= \langle Y, T\rangle|_a^b - \int_a^b \left\langle Y, \frac{dT}{ds} \right\rangle ds$$

$$\frac{d}{dt}\Big|_{t=0}\mathcal{B}(\gamma_t) = \int_a^b \left(\frac{1}{2}\left\langle \frac{dT}{ds}, \frac{dT}{ds} \right\rangle ds \right)^{\cdot}$$

$$= \int_a^b \left(\left\langle \frac{d^2Y}{ds^2}, \frac{dT}{ds} \right\rangle - \frac{3}{2}\left\langle \frac{dY}{ds}, T \right\rangle\left\langle \frac{dT}{ds}, \frac{dT}{ds} \right\rangle \right) ds$$

$$= \int_a^b \left(\frac{d}{ds}\left\langle \frac{dY}{ds}, \frac{dT}{ds} \right\rangle - \left\langle \frac{dY}{ds}, \frac{d^2T}{ds^2} \right\rangle - \frac{3}{2}\frac{d}{ds}\left(\langle Y, T\rangle\left\langle \frac{dT}{ds}, \frac{dT}{ds} \right\rangle \right) \right.$$

$$\left. + \frac{3}{2}\left\langle Y, \frac{dT}{ds} \right\rangle\left\langle \frac{dT}{ds}, \frac{dT}{ds} \right\rangle + 3\langle Y, T\rangle\left\langle \frac{dT}{ds}, \frac{d^2T}{ds^2} \right\rangle \right) ds$$

$$= \int_a^b \frac{d}{ds}\left(\left\langle \frac{dY}{ds}, \frac{dT}{ds} \right\rangle - \left\langle Y, \frac{d^2T}{ds^2} \right\rangle - \frac{3}{2}\langle Y, T\rangle\left\langle \frac{dT}{ds}, \frac{dT}{ds} \right\rangle \right) ds$$

$$+ \int_a^b \left(\left\langle Y, \frac{d^3T}{ds^3} + 3\left\langle \frac{dT}{ds}, \frac{d^2T}{ds^2} \right\rangle T + \frac{3}{2}\left\langle \frac{dT}{ds}, \frac{dT}{ds} \right\rangle\frac{dT}{ds} \right\rangle \right) ds$$

$$= \left(\left\langle \frac{dY}{ds}, \frac{dT}{ds} \right\rangle - \left\langle Y, \frac{d^2T}{ds^2} + \frac{3}{2}\left\langle \frac{dT}{ds}, \frac{dT}{ds} \right\rangle T \right\rangle \right)\Big|_a^b$$

$$+ \int_a^b \left(\left\langle Y, \frac{d^3T}{ds^3} + 3\left\langle \frac{dT}{ds}, \frac{d^2T}{ds^2} \right\rangle T + \frac{3}{2}\left\langle \frac{dT}{ds}, \frac{dT}{ds} \right\rangle\frac{dT}{ds} \right\rangle \right) ds.$$

□

2.3   Critical Points of Length and Bending Energy

Variations of curves (as defined in Definition 2.3) are needed in order to define and determine those curves that represent equilibria of geometrically interesting variational functionals. Functionals are certain real-valued functions on the space $\mathcal{M}$ of all curves $\gamma : [a, b] \to \mathbb{R}^n$, that was already introduced in Sect. 2.1.

Definition 2.10

Suppose we have a way to assign to each curve $\gamma : [a, b] \to \mathbb{R}^n$ a real number $\mathcal{E}(\gamma)$. Then $\mathcal{E}$ is called a **smooth functional** if for every smooth one-parameter family

$$t \mapsto \gamma_t, \quad t \in [t_0, t_1]$$

of curves $\gamma_t : [a, b] \to \mathbb{R}^n$ the function

$$[t_0, t_1] \to \mathbb{R}, \ t \mapsto \mathcal{E}(\gamma_t)$$

is smooth.

In many circumstances, we want to consider only variations of $\gamma : [a, b] \to \mathbb{R}^n$ that keep the curve fixed near the boundary of the interval $[a, b]$ (see Fig. 2.2).

Definition 2.11

Let $\gamma : [a, b] \to \mathbb{R}^n$ a curve. Then a variation

$$t \mapsto \gamma_t, \quad t \in [t_0, t_1]$$

of γ is said to have **support in the interior** of $[a, b]$ if there is $\epsilon > 0$ such that for all $x \in [a, a + \epsilon] \cup [b - \epsilon, b]$ we have

$$\gamma_t(x) = \gamma(x) \quad \text{for all} \quad t \in [t_0, t_1].$$

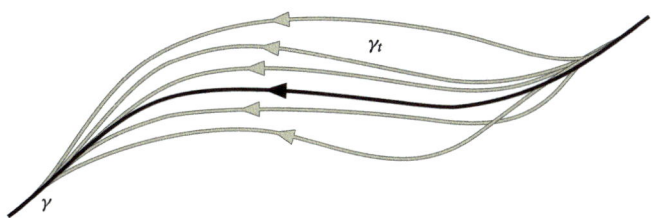

Fig. 2.2 A variation γ_t of a curve γ with support in the interior

Now we can make precise what we meant by an equilibrium of a variational energy:

Definition 2.12

Let $\mathcal{E}$ be a smooth functional on the space of curves $\gamma: [a, b] \to \mathbb{R}^n$. Then a curve $\gamma: [a, b] \to \mathbb{R}^n$ is called a **critical point** of $\mathcal{E}$ if for all variations $t \mapsto \gamma_t$ of γ with support in the interior of $[a, b]$ we have

$$\left.\frac{d}{dt}\right|_{t=0} \mathcal{E}(\gamma_t) = 0.$$

We denote the space

$$\left\{ Y: [a, b] \to \mathbb{R}^n \text{ smooth} \;\middle|\; Y|_{[a,a+\delta] \cup [b-\delta, b]} = 0 \text{ for some } \delta > 0 \right\}$$

of all functions $Y: [a, b] \to \mathbb{R}^n$ with support in the interior of $[a, b]$ (Definition A.4) by $C_0^\infty((a, b), \mathbb{R}^n)$.

Theorem 2.13
For every vector field $Y: [a, b] \to \mathbb{R}^n$ along a curve $\gamma: [a, b] \to \mathbb{R}^n$ there is a variation $t \mapsto \gamma_t$ with variational vector field Y. If Y has support in the interior of $[a, b]$, then also the variation $t \mapsto \gamma_t$ can be chosen in such a way that it has support in the interior of $[a, b]$.

Proof. The proof of Theorem 2.13 is left as an exercise.   □

▶ **Remark 2.14** In the case of the length functional, instead of using variations with support in the interior we could have used variations that fix both end points. For other variational problems (that involve higher derivative), additional derivatives (not only the function value) of γ would have to be clamped to fixed values at the end points. On the other hand, variations with support in the interior will work all the time, with equivalent results.

Theorem 2.15 (Fundamental Lemma of the Calculus of Variations)
On the vector space $C^\infty([a, b], \mathbb{R}^n)$ equipped with the inner product

$$\langle\!\langle f, g \rangle\!\rangle := \int_a^b \langle f, g \rangle$$

(continued)

Theorem 2.15 (continued)
only the zero vector is in the orthogonal complement of $C_0^\infty ((a, b), \mathbb{R}^n)$:

$$C_0^\infty ((a, b), \mathbb{R}^n)^\perp = \{0\}.$$

Proof. Suppose that $f \in C^\infty ([a, b], \mathbb{R}^n)$ would be non-zero but in the orthogonal complement $C_0^\infty ((a, b), \mathbb{R}^n)^\perp$. Then there would be $x_0 \in [a, b]$ such that $f(x_0) \neq 0$. Choose $\delta > 0$ such that $[x_0 - \delta, x_0 + \delta] \subset (a, b)$ and $\langle f(x), f(x_0) \rangle > 0$ for all $x \in [x_0 - \delta, x_0 + \delta]$. Construct a smooth bump function (cf. Appendix A.2)

$$g \in C_0^\infty ((x_0 - \delta, x_0 + \delta), \mathbb{R}^n) \subset C_0^\infty ((a, b), \mathbb{R}^n)$$

such that $g \geq 0$ and $g(x_0) = 1$. Then $\langle f, g \rangle \neq 0$, which implies $f \notin C_0^\infty ((a, b), \mathbb{R}^n)^\perp$, a contradiction. $\square$

Now we are in the position to determine the critical points of the length functional:

Theorem 2.16
A curve $\gamma : [a, b] \to \mathbb{R}^n$ is a critical point of the length functional $\mathcal{L}$ if and only if its unit tangent field $T : [a, b] \to \mathbb{R}^n$ is constant, i.e. if γ parametrizes a straight line segment.

Proof. By Theorem 2.9 and 2.15, γ is a critical point of $\mathcal{L}$ if and only if for all $Y \in C_0^\infty ((a, b), \mathbb{R}^n)$ we have

$$\left\langle \left\langle Y, \frac{dT}{ds} \right\rangle \right\rangle = 0.$$

By Theorem 2.15 this is the case if and only if $\frac{dT}{ds} = 0$. $\square$

Definition 2.17

A curve $\gamma : [a, b] \to \mathbb{R}^n$ is called a **free elastic curve** if it is a critical point of the bending energy functional $\mathcal{B}$.

An almost identical proof as the one of Theorem 2.16 gives us

Theorem 2.18

A curve $\gamma : [a, b] \to \mathbb{R}^n$ is a free elastic curve if and only if its unit tangent field $T : [a, b] \to \mathbb{R}^n$ satisfies

$$\frac{d^3T}{ds^3} + 3\left\langle \frac{dT}{ds}, \frac{d^2T}{ds^2} \right\rangle T + \frac{3}{2}\left\langle \frac{dT}{ds}, \frac{dT}{ds} \right\rangle \frac{dT}{ds} = 0$$

or equivalently

$$\frac{d^4\gamma}{ds^3} + 3\left\langle \frac{d^2\gamma}{ds^2}, \frac{d^3\gamma}{ds^3} \right\rangle \frac{d\gamma}{ds} + \frac{3}{2}\left\langle \frac{d^2\gamma}{ds^2}, \frac{d^2\gamma}{ds^2} \right\rangle \frac{d^2\gamma}{ds^2} = 0.$$

Solving the fourth order differential equation for γ that appears in Theorem 2.18 with suitable initial values will give us unit speed parametrizations of free elastic curves. In the next chapter we will explore in more detail the geometric consequences of this differential equation.

2.4 Constrained Variation

In the context of many variational problems that arise in applications, general variations might violate some constraints that are imposed by the nature of the problem at hand. For example, thin elastic wires (for most practical purposes) have a fixed length. This means that here we should minimize bending energy only among those curves (held fixed near their boundary) that have a prescribed length.

This kind of problem is known under the name of **optimization under constraints**. Here we will work with a definition of a critical point under constraints that is slightly stronger than the standard one. The usual definition would replace the condition $\frac{d}{dt}\big|_{t=0} \tilde{\mathcal{E}} = 0$ by the requirement that $\tilde{\mathcal{E}}(\gamma_t)$ is independent of t.

Definition 2.19

Let $\mathcal{E}, \tilde{\mathcal{E}}$ be two smooth functionals on the space of all curves $\tilde{\gamma} : [a, b] \to \mathbb{R}^n$. Then a curve $\gamma : [a, b] \to \mathbb{R}^n$ is called a **critical point of $\mathcal{E}$ under the constraint** of fixed $\tilde{\mathcal{E}}$, if for all variations $t \mapsto \gamma_t$ of γ with support in the interior of $[a, b]$

$$\frac{d}{dt}\bigg|_{t=0} \tilde{\mathcal{E}} = 0$$

implies

$$\frac{d}{dt}\Big|_{t=0} \mathcal{E} = 0.$$

Both for the length functional $\mathcal{E} = \mathcal{L}$ and for the bending energy $\mathcal{E} = \mathcal{B}$ we know (Theorems 2.8 and 2.9) how to express the infinitesimal variation of $\mathcal{E}$ that corresponds to a variational vector field $Y : [a, b] \to \mathbb{R}^n$ with support in the interior of $[a, b]$ as an integral

$$\frac{d}{dt}\Big|_{t=0} \mathcal{E} = \int_a^b \langle Y, G_\gamma \rangle$$

for some smooth map $G_\gamma : [a, b] \to \mathbb{R}^n$. If a formula like the one above holds, G_γ is called the **gradient** of the energy $\mathcal{E}$ at γ.

Theorem 2.20

Let $\mathcal{E}, \tilde{\mathcal{E}}$ be two smooth functionals on the space of all curves $\tilde{\gamma} : [a, b] \to \mathbb{R}^n$. Suppose we have a way to associate to each curve $\gamma : [a, b] \to \mathbb{R}^n$ smooth maps

$$G_\gamma, \tilde{G}_\gamma : [a, b] \to \mathbb{R}^n$$

such that for all variations $t \mapsto \gamma_t$ of γ with support in the interior of $[a, b]$ we have

$$\frac{d}{dt}\Big|_{t=0} \mathcal{E} = \int_a^b \langle \dot{\gamma}, G_\gamma \rangle$$

$$\frac{d}{dt}\Big|_{t=0} \tilde{\mathcal{E}} = \int_a^b \langle \dot{\gamma}, \tilde{G}_\gamma \rangle.$$

Then γ is a critical point of $\mathcal{E}$ under the constraint of fixed $\tilde{\mathcal{E}}$ if and only if there is a constant $\lambda \in \mathbb{R}$ such that

$$G_\gamma = \lambda \tilde{G}_\gamma.$$

*λ is called a **Lagrange multiplier** for the constraint of fixed $\tilde{\mathcal{E}}$.*

Proof. We apply Theorem 2.21 below to the case where $H = C^\infty([a, b], \mathbb{R}^n)$, $V = C_0^\infty((a, b), \mathbb{R}^n)$ and $U = \mathbb{R}\tilde{G}_\gamma$. Then γ is a critical point of $\mathcal{E}$ under the constraint of fixed $\tilde{\mathcal{E}}$ if and only if G_γ is orthogonal to all Y that are simultaneously in V and orthogonal to U, i.e

$$G_\gamma \in (V \cap U^\perp)^\perp = U = \mathbb{R}\tilde{G}_\gamma.$$

$\square$

The theorem below is pure linear algebra, no Functional Analysis is involved. The formulation is such that it can also be applied to a situation where there are constraint functionals $\mathcal{E}_1, \ldots, \mathcal{E}_k$ instead of a single functional $\tilde{\mathcal{E}}$.

Theorem 2.21

Let H be a (possibly infinite dimensional) vector space with inner product $\langle ., . \rangle$. Let $V \subset H$ be a subspace such that $V^\perp = \{0\}$ and $U \subset H$ finite dimensional. Then

$$(U^\perp \cap V)^\perp = U.$$

Proof. As for all $x \in U^\perp \cap V$ it holds that $\langle u, x \rangle = 0$ for all $u \in U$, the inclusion $U \subset (U^\perp \cap V)^\perp$ is immediate. In order to show that also $(U^\perp \cap V)^\perp \subset U$ we choose an orthonormal basis $\{u_1, \ldots, u_n\}$ of U and define the map

$$P : H \to U, \ x \mapsto \sum_{i=1}^n \langle x, u_i \rangle u_i.$$

It is not hard to check that P defines an orthogonal projection of H onto U, i.e. $P^2 = P$, $P^* = P$ and $\operatorname{im} P = U$. Now for $u \in U$ and $h \in H$ it holds

$$\langle u, h \rangle = \langle P(u), h \rangle = \langle u, P(h) \rangle.$$

Therefore we have $U \cap P(V)^\perp \subset V^\perp = \{0\}$, hence $P(V) = U$. So there are $v_1, \ldots, v_n \in V$ such that $P(v_i) = u_i$. We now define the map

$$Q : H \to V, \ x \mapsto \sum_{i=1}^n \langle x, v_i \rangle v_i$$

which is symmetric (i.e. $Q^* = Q$) and satisfies $\operatorname{im} Q \subset V$ and $P \circ Q|_U = \operatorname{id}_U$. Therefore, for $x \in (U^\perp \cap V)^\perp$ and $v \in V$:

$$\langle x - P \circ Q(x), v \rangle = \langle x, v - Q \circ P(v) \rangle = 0,$$

Fig. 2.3 Elastic curves are everywhere

since $v - Q \circ P(v) \in U^\perp \cap V$. Thus $x - P \circ Q(x) \in V^\perp = \{0\}$ and therefore $x = P \circ Q(x) \in U$.   □

Definition 2.22

A curve $\gamma : [a, b] \to \mathbb{R}^n$ is called a **torsion-free elastic curve** if it is a critical point of bending energy under the constraint of fixed length (Fig. 2.3).

Theorems 2.8, 2.9 and 2.20 together allow us to characterize torsion-free elastic curves by a differential equation:

Theorem 2.23

A curve $\gamma : [a, b] \to \mathbb{R}^n$ is a torsion-free elastic curve if and only if there is a constant $\lambda \in \mathbb{R}$ such that its unit tangent field satisfies

$$\frac{d^3 T}{ds^3} + 3\left\langle \frac{dT}{ds}, \frac{d^2 T}{ds^2} \right\rangle T + \frac{3}{2}\left\langle \frac{dT}{ds}, \frac{dT}{ds} \right\rangle \frac{dT}{ds} - \lambda \frac{dT}{ds} = 0$$

or, equivalently,

$$\frac{d^4 \gamma}{ds^4} + 3\left\langle \frac{d^2 \gamma}{ds^2}, \frac{d^3 \gamma}{ds^3} \right\rangle \frac{d\gamma}{ds} + \frac{3}{2}\left\langle \frac{d^2 \gamma}{ds^2}, \frac{d^2 \gamma}{ds^2} \right\rangle \frac{d^2 \gamma}{ds^2} - \lambda \frac{d^2 \gamma}{ds^2} = 0.$$

*The constant λ is called the **tension** of γ.*

2.5　Torsion-Free Elastic Curves and the Pendulum Equation

By Theorem 1.16 every curve in $\mathbb{R}^n$ admits a reparametrization $\gamma \colon [0, L] \to \mathbb{R}^n$ with unit speed. Then for any function $g \colon [0, L] \to \mathbb{R}^k$ the derivative with respect to arclength is just the ordinary derivative:

Theorem 2.24

A curve $\gamma \colon [0, L] \to \mathbb{R}^n$ with unit speed is torsion-free elastic with tension λ if and only if its unit tangent field $T \colon [a, b] \to S^{n-1}$ solves the equation of motion

$$T'' - \langle T'', T \rangle T = \mathbf{a} - \langle \mathbf{a}, T \rangle T$$

*of a spherical pendulum with unit mass and some **gravity vector** $\mathbf{a} \in \mathbb{R}^n$ and λ equals the total energy of the pendulum:*

$$\lambda = \frac{1}{2} \langle T', T' \rangle - \langle \mathbf{a}, T \rangle.$$

Proof. Let $\gamma \colon [0, L] \to \mathbb{R}^n$ be a torsion-free elastic curve with tension λ and with unit speed. Then, by Theorem 2.23

$$0 = T''' + 3\langle T', T'' \rangle T + \frac{3}{2} \langle T', T' \rangle T' - \lambda T' = \left(T'' + \frac{3}{2} \langle T', T' \rangle T - \lambda T \right)',$$

i.e. if there is a constant vector $\mathbf{a} \in \mathbb{R}^n$ such that

$$T'' + \frac{3}{2} \langle T', T' \rangle T - \lambda T = \mathbf{a}.$$

Looking at the component orthogonal to T on both sides of this equation gives us the first of the two equations that we want to prove. Taking the scalar product with T and using

$$0 = \frac{1}{2} \langle T, T \rangle'' = \langle T'', T \rangle + \langle T', T' \rangle$$

we obtain the second equation. Conversely, if $T \colon [0, L] \to S^{n-1}$ solves the pendulum equation

$$T'' - \langle T'', T \rangle T = \mathbf{a} - \langle \mathbf{a}, T \rangle T,$$

then it is easy to verify that the total energy λ defined by

$$\lambda = \frac{1}{2}\langle T', T'\rangle - \langle \mathbf{a}, T\rangle$$

is constant and

$$T'' + \frac{3}{2}\langle T', T'\rangle T - \lambda T = \mathbf{a}.$$

$\square$

Figure 2.4 shows planar torsion-free elastic curves that lie in a plane. They arise from pendulum motion on a circle, whereas a pendulum motion on S^2 gives a torsion-free elastic curve in $\mathbb{R}^3$ as seen in Fig. 2.5.

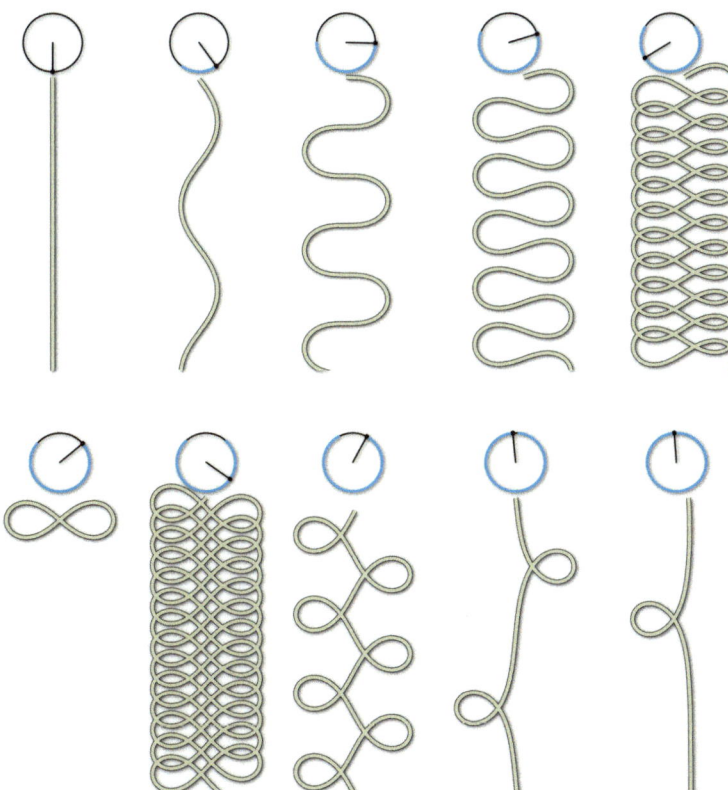

Fig. 2.4 Trajectories of a pendulum (drawn in blue color). Below each of these trajectories the corresponding torsion-free elastic curve is shown

Fig. 2.5 Trajectory of a pendulum on S^2 (drawn in blue color), together with the corresponding torsion-free elastic curve in $\mathbb{R}^3$

Curves in $\mathbb{R}^2$

3

Curves in the plane $\mathbb{R}^2$ are special in several respects: For a closed plane curve γ an enclosed area $\mathcal{A}(\gamma)$ can be defined, providing another geometric functional in addition to length and bending energy. Unlike the situation in higher dimensions, the geometry of an arbitrary unit speed plane curve $\gamma : [a, b] \to \mathbb{R}^2$ is captured in a smooth real-valued curvature function $\kappa : [a, b] \to \mathbb{R}$. We prove our first theorem in Global Differential Geometry: The integral of the curvature of a closed plane curve is $2\pi n$ where n is an integer, called the *tangent winding number* of γ. Two closed plane curves can be smoothly deformed into each other if and only if they have the same tangent winding number.

3.1 Plane Curves

The case of curves $\gamma : [a, b] \to \mathbb{R}^2$ is special because $\mathbb{R}^2$ comes with a distinguished linear map $J : \mathbb{R}^2 \to \mathbb{R}^2$, the 90°-**rotation** in the counterclockwise (positive) direction:

$$J : \mathbb{R}^2 \to \mathbb{R}^2, \qquad J\begin{pmatrix} x \\ y \end{pmatrix} = \begin{pmatrix} 0 & -1 \\ 1 & 0 \end{pmatrix} \begin{pmatrix} x \\ y \end{pmatrix}$$

Here are some properties of J that are easy to check: We have $J^2 = -I$ and J is orthogonal as well as skew-adjoint, i.e. for all vectors $X, Y \in \mathbb{R}^2$ we have

$$\langle JX, JY \rangle = \langle X, Y \rangle$$
$$\langle JX, Y \rangle = -\langle X, JY \rangle.$$

© The Author(s) 2024
U. Pinkall, O. Gross, *Differential Geometry*, Compact Textbooks in Mathematics,
https://doi.org/10.1007/978-3-031-39838-4_3

Furthermore, the determinant function det on $\mathbb{R}^2$ can be expressed in terms of J and the scalar product:

$$\langle JX, Y \rangle = \det(X, Y).$$

If $\gamma : [a, b] \to \mathbb{R}^2$ is a curve and $T : [a, b] \to \mathbb{R}^2$ is its unit tangent, then $\frac{dT}{ds}$ is orthogonal to T and therefore proportional to JT:

Definition 3.1

Let $\gamma : [a, b] \to \mathbb{R}^2$ be a curve and $T : [a, b] \to \mathbb{R}^2$ its unit tangent. Then the unique function $\kappa : [a, b] \to \mathbb{R}$ such that

$$\frac{dT}{ds} = \kappa J T$$

is called the **curvature** of γ.

More explicitly,

$$\kappa = \left\langle JT, \frac{dT}{ds} \right\rangle = \left\langle \frac{1}{v} J\gamma', \frac{1}{v} \left(\frac{1}{v} \gamma' \right)' \right\rangle = \frac{\det(\gamma', \gamma'')}{|\gamma'|^3}.$$

We convince ourselves that the curvature is independent of the parametrization: Let $\tilde{\gamma} = \gamma \circ \varphi$ be a reparametrization of a plane curve γ. Then

$$\tilde{\gamma}' = \varphi' \cdot \gamma' \circ \varphi$$

and, since $\varphi' > 0$, $|\tilde{\gamma}'| = \varphi' |\gamma' \circ \varphi|$. Hence

$$\tilde{\gamma}'' = \varphi'' \cdot \gamma' \circ \varphi + \varphi'(\varphi' \cdot \gamma'' \circ \varphi)$$

and therefore

$$\tilde{\kappa} = \frac{\det(\tilde{\gamma}', \tilde{\gamma}'')}{|\tilde{\gamma}'|^3} = \frac{(\varphi')^3 \cdot \det(\gamma', \gamma'')}{(\varphi')^3 |\gamma'|^3} = \kappa .$$

The curvature κ of a straight line segment vanishes since we can parametrize the segment going from a point $p \in \mathbb{R}^2$ to a point $q \in \mathbb{R}^2$ by

$$\gamma : [a, b] \to \mathbb{R}^2, \ x \mapsto p + \frac{x - a}{b - a}(q - p)$$

and therefore $\gamma'' = 0$, so that $\kappa = 0$. A circular arc

$$\gamma: [a, b] \to \mathbb{R}^2, \ x \mapsto \begin{pmatrix} r\cos x \\ r\sin x \end{pmatrix}$$

of radius r has constant curvature $\kappa = \frac{1}{r}$. If we restrict attention to unit speed curves $\gamma: [0, L] \to \mathbb{R}^2$, the curvature function $\kappa: [0, L] \to \mathbb{R}$ determines γ up to orientation-preserving congruence:

Theorem 3.2 (Fundamental Theorem of Plane Curves)

(i) *For every smooth function* $\kappa: [0, L] \to \mathbb{R}$ *there is a unit speed curve* $\gamma: [0, L] \to \mathbb{R}^2$ *with curvature* κ.

(ii) *If* $\gamma, \tilde{\gamma}: [0, L] \to \mathbb{R}^2$ *are unit speed curves with the same curvature function* κ, *then there is an orthogonal* (2×2)*-matrix* A *with* $\det A = 1$ *and a vector* $\mathbf{b} \in \mathbb{R}^2$ *such that*

$$\tilde{\gamma} = A\gamma + \mathbf{b}.$$

Proof. For (ii), denote by T, $\tilde{T}$ the unit tangent fields of γ and $\tilde{\gamma}$ and take for A the orthogonal (2×2)-matrix A with determinant one for which $AT(0) = \tilde{T}(0)$. Then both $\tilde{T}$ and

$$\hat{T} := AT$$

solve the linear initial value problem

$$Y(0) = \tilde{T}(0)$$
$$Y' = \kappa J Y$$

and therefore, by the uniqueness part of the Picard-Lindelöf theorem, we must have $\hat{T} = \tilde{T}$. Then

$$(\tilde{\gamma} - A\gamma)' = \kappa J (\tilde{T} - \hat{T}) = 0,$$

which proves (ii). For (i), define $\alpha: [0, L] \to \mathbb{R}$ and $T, \gamma: [0, L] \to \mathbb{R}^2$ by

$$\alpha(x) := \int_0^x \kappa$$

$$T := \begin{pmatrix} \cos\alpha \\ \sin\alpha \end{pmatrix}$$

$$\gamma(x) := \int_0^x T.$$

Then $|T| = 1$ and $\gamma' = T$, so γ is a curve and T is its unit tangent field. Furthermore, $T' = \kappa J T$ and therefore γ has curvature κ.   □

3.2   Area of a Plane Curve

Let $\gamma: [a, b] \to \mathbb{R}^2$ be a curve such that $\det(\gamma, \gamma') > 0$ and the map

$$f: (0, 1] \times [a, b] \to \mathbb{R}^2, \quad f(t, x) = t\gamma(x)$$

is a bijective map onto a subset $S \subset \mathbb{R}^2$. Then the derivative $f'(t, x)$ at the point $(t, x) \in (0, 1) \times [a, b]$ satisfies

$$\det f'(t, x) = t \det(\gamma(x), \gamma'(x)) > 0$$

and using the transformation formula of integrals it is not difficult to show that the area of S is given by

$$\text{area}(S) = \int_S 1 = \int_{f((0,1]\times[a,b])} 1 = \int_a^b \int_0^1 \det f' = \frac{1}{2}\int_a^b \det(\gamma, \gamma').$$

For the curve γ shown on the left of Fig. 3.1, the above formula correctly yields the area enclosed by γ and the line segments from the origin to $\gamma(a)$ and $\gamma(b)$. It

Fig. 3.1 For the curve on the right, the position vector from the origin to $\gamma(x)$ covers some areas multiple times

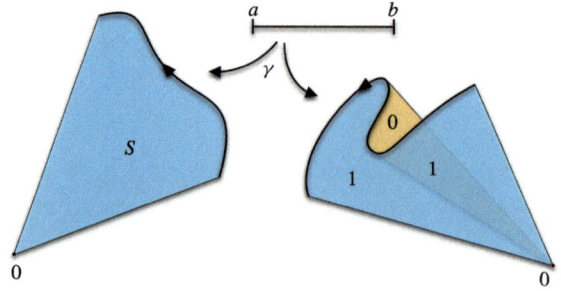

therefore seems reasonable to use this formula in order to define an area for arbitrary curves $\gamma : [a, b] \rightarrow \mathbb{R}^2$:

Definition 3.3

The **sector area** of a curve $\gamma : [a, b] \rightarrow \mathbb{R}^2$ is defined as

$$\mathcal{A}(\gamma) = \frac{1}{2} \int_a^b \det(\gamma, \gamma').$$

The curve on the right of Fig. 3.1 illustrates the consequences of this definition. There the position vector from the origin to $\gamma(x)$ covers some areas multiple times. However, for some of these times (where γ, as seen from the origin, moves clockwise) the contribution to the covered area, as it is computed by the above formula, is negative.

The sector area $\mathcal{A}(\gamma)$ depends on the origin in $\mathbb{R}^2$, which means that it changes if we apply a translation $\mathbf{p} \mapsto \mathbf{p} - \mathbf{v}$ to γ. Therefore, at first sight the sector area does not look like a good geometric invariant for curves. However, this dependence disappears as soon as we restrict attention to closed curves, or consider differences between the sector areas of curves that share the same endpoints (see Fig. 3.2):

Let $\mathbf{v} \in \mathbb{R}^2$ be a vector and $\gamma : [a, b] \rightarrow \mathbb{R}^2$ a curve. Then we define a modified sector area $\mathcal{A}_{\mathbf{v}}(\gamma)$ as the sector area of the curve γ translated by the vector $\mathbf{v}$:

$$\mathcal{A}_{\mathbf{v}}(\gamma) := \mathcal{A}(\gamma + \mathbf{v}).$$

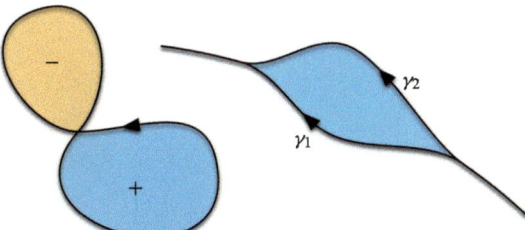

Fig. 3.2 Independent of the choice of origin in $\mathbb{R}^2$, the sector area of the curve on the left of the above picture equals the area of the blue region minus the area of the orange region. Similarly, the difference of the sector areas of the two curves on the right equals the area of the blue region between them

> **Theorem 3.4**
>
> Let $\gamma : [a, b] \to \mathbb{R}^2$ be a closed curve and $\gamma_1, \gamma_2 : [a, b] \to \mathbb{R}^2$ two curves
> with $\gamma_1(a) = \gamma_2(a)$ and $\gamma_1(b) = \gamma_2(b)$. Then, for any vector $\mathbf{v} \in \mathbb{R}^2$ we have
>
> $$\mathcal{A}_{\mathbf{v}}(\gamma) = \mathcal{A}(\gamma)$$
>
> $$\mathcal{A}_{\mathbf{v}}(\gamma_2) - \mathcal{A}_{\mathbf{v}}(\gamma_1) = \mathcal{A}(\gamma_2) - \mathcal{A}(\gamma_1).$$

Proof. Because γ is closed, we have

$$
\begin{aligned}
\mathcal{A}_{\mathbf{v}}(\gamma) - \mathcal{A}(\gamma) &= \frac{1}{2} \int_a^b \det(\gamma + \mathbf{v}, \gamma') - \frac{1}{2} \int_a^b \det(\gamma, \gamma') \\
&= \frac{1}{2} \int_a^b \det(\mathbf{v}, \gamma') \\
&= \frac{1}{2} \int_a^b \det(\mathbf{v}, \gamma)' \\
&= \frac{1}{2} \det(\mathbf{v}, \gamma)\big|_a^b \\
&= 0.
\end{aligned}
$$

By the same arguments we obtain

$$
\begin{aligned}
(\mathcal{A}_{\mathbf{v}}(\gamma_2) &- \mathcal{A}_{\mathbf{v}}(\gamma_1)) - (\mathcal{A}(\gamma_2) - \mathcal{A}(\gamma_1)) \\
&= \frac{1}{2} \int_a^b \det(\mathbf{v}, \gamma_2)' - \frac{1}{2} \int_a^b \det(\mathbf{v}, \gamma_1)' \\
&= \det(\mathbf{v}, \gamma_2 - \gamma_1)\big|_a^b \\
&= 0.
\end{aligned}
$$

$\square$

In particular, we expect that for variations with support in the interior of $[a, b]$ of
a curve $\gamma : [a, b] \to \mathbb{R}^2$, the corresponding variation of sector area is independent
of the choice of origin:

Theorem 3.5
Let $t \mapsto \gamma_t$ be a variation with support in the interior of $[a, b]$ of a curve $\gamma : [a, b] \to \mathbb{R}^2$. Then

$$\frac{d}{dt}\Big|_{t=0} \mathcal{A}(\gamma_t) = -\int_a^b \langle Y, J\gamma' \rangle.$$

Proof. Since Y vanishes at the endpoints, we have

$$
\begin{aligned}
\frac{d}{dt}\Big|_{t=0} \mathcal{A}(\gamma_t) &= \frac{1}{2}\int_a^b \det(Y, \gamma') + \frac{1}{2}\int_a^b \det(\gamma, (\gamma')^{\boldsymbol{\cdot}}) \\
&= \frac{1}{2}\int_a^b \det(Y, \gamma') + \frac{1}{2}\int_a^b \det(\gamma, Y') \\
&= \frac{1}{2}\int_a^b \det(Y, \gamma') - \frac{1}{2}\int_a^b \det(\gamma', Y) \\
&= \int_a^b \det(Y, \gamma') \\
&= -\int_a^b \langle Y, J\gamma' \rangle.
\end{aligned}
$$

$\square$

As a consequence, the sector area functional by itself does not have any critical points. On the other hand, minimizing length among all curves with the same endpoints and the same sector area is possible:

Theorem 3.6
A curve $\gamma : [a, b] \to \mathbb{R}^2$ is a critical point of length under the constraint of fixed sector area if and only if its curvature κ is constant, i.e. if and only if its image lies on a circle or a straight line.

Proof. By Theorems 2.9 and 2.20 γ is a critical point of length under the constraint of fixed sector area if and only if there is a constant $\lambda \in \mathbb{R}$ such that

$$\lambda(-J\gamma') = -T' = -\kappa J\gamma'.$$

$\square$

3.3   Planar Elastic Curves

For a unit speed curve $\gamma : [0, L] \to \mathbb{R}^2$ with unit tangent T and curvature κ we have

$$T' = \kappa J T$$

$$T'' = -\kappa^2 T + \kappa' J T$$

$$T''' = -3\kappa\kappa' T + (\kappa'' - \kappa^3) J T$$

The bending energy of a plane curve is also called its **total squared curvature**. This is because for a unit speed plane curve γ as above we have

$$\mathcal{B}(\gamma) = \frac{1}{2} \int_a^b \langle T', T' \rangle \, ds = \frac{1}{2} \int_a^b \kappa^2 \, ds.$$

By Theorem 2.23, γ is an elastic curve with tension λ if and only if

$$0 = T''' + 3 \langle T', T'' \rangle T + \frac{3}{2} \langle T', T' \rangle T' - \lambda T'$$

$$= \left(\kappa'' + \frac{\kappa^3}{2} + \lambda\kappa \right) J T$$

which means

$$\kappa'' + \frac{\kappa^3}{2} + \lambda\kappa = 0.$$

This differential equation can be interpreted as the equation of motion

$$\kappa'' + \frac{\partial V}{\partial \kappa}(\kappa) = 0$$

for a particle with unit mass moving on the real line subject to the potential energy

$$V(\kappa) = \tfrac{1}{8}\kappa^4 + \tfrac{\lambda}{2}\kappa^2.$$

As expected (and as is easy to verify by taking the derivative) the total energy

$$E := \frac{1}{2}(\kappa')^2 + V(\kappa)$$

is constant. In particular, we see that along for each solution the potential energy is bounded from above by E. In Fig. 3.3 we see examples that should be compared to the shapes of the corresponding curves that were shown in Sect. 2.5.

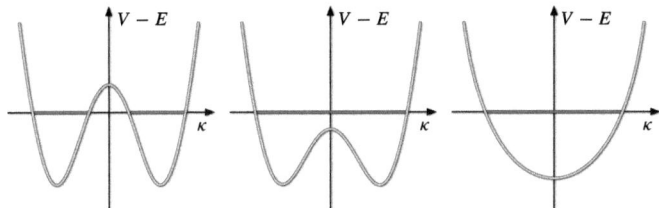

Fig. 3.3 The potential wells for various values of λ. For a solution κ of the equation of motion, $V(\kappa) - E$ is always non-positive. The values of κ that satisfy this condition are indicated in blue

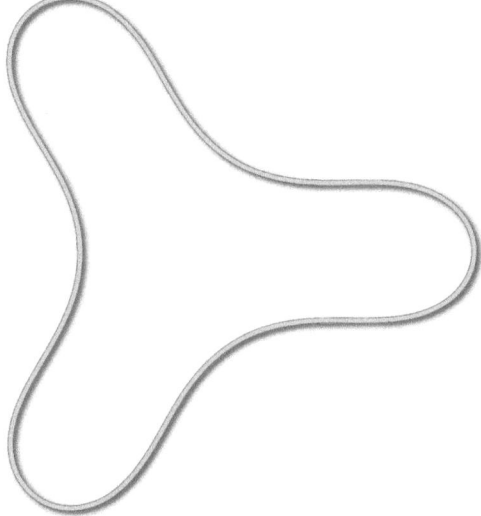

Fig. 3.4 A curve which is a critical points of the total squared curvature with constrained length and the sector area

If we look for critical points of the total squared curvature while constraining not only the length but also the sector area, by Theorem 3.6 we arrive at the differential equation for κ:

$$\kappa'' + \frac{\kappa^3}{2} + \lambda\kappa + \mu = 0.$$

The closed curve in Fig. 3.4 is such a critical point.

3.4   Tangent Winding Number

Definition 3.7

For a curve $\gamma : [a, b] \to \mathbb{R}^2$ with curvature κ the integral

$$\int_a^b \kappa \, ds$$

is called the **total curvature** of γ.

In this section we will prove that for a closed curve in $\mathbb{R}^2$ the total curvature is an integer multiple of 2π:

Theorem 3.8
If $\gamma : [a, b] \to \mathbb{R}^2$ is a closed curve with curvature κ, then there is an integer $n \in \mathbb{Z}$ such that

$$\int_a^b \kappa \, ds = 2\pi n.$$

*n is called the **tangent winding number** of γ.*

Proof. Define $\alpha : [a, b] \to \mathbb{R}$ by

$$\alpha(x) := \alpha_0 + \int_a^x \kappa \, ds$$

where α_0 is chosen in such a way that

$$T(a) = (\cos\alpha_0, \sin\alpha_0).$$

As in the proof of Theorem 3.2, we conclude

$$T = (\cos\alpha, \sin\alpha).$$

Since γ is closed, we have $T(b) = T(a)$, which means

$$(\cos\alpha(b), \sin\alpha(b)) = (\cos\alpha(a), \sin\alpha(a)).$$

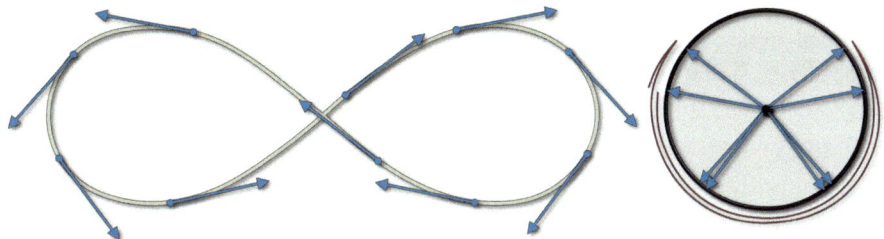

Fig. 3.5 The path on S^1 of the unit tangent can be visualized more clearly if it is drawn slightly outside of the unit circle

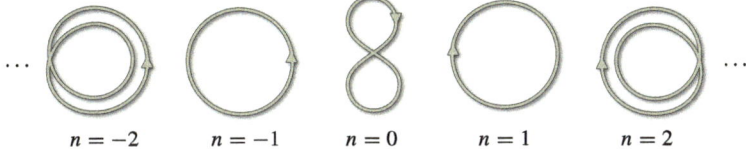

$$n = -2 \qquad n = -1 \qquad n = 0 \qquad n = 1 \qquad n = 2$$

Fig. 3.6 A list of representatives for every homotopy class of plane curves

Therefore, there is an integer $n \in \mathbb{Z}$ such that

$$\int_a^b \kappa \, ds = \alpha(b) - \alpha(a) = 2\pi n.$$

$\square$

As is clear from the above proof, the tangent winding number counts how often the unit tangent $T(x)$ turns around the unit circle S^1 as x runs from a to b (see Fig. 3.5). Figure 3.6 shows that all integers $n \in \mathbb{Z}$ arise as the tangent winding number of some curve in $\mathbb{R}^2$.

3.5 Regular Homotopy

The following two sections will deal with the question: "Given two curves $\gamma, \tilde{\gamma}$ in $\mathbb{R}^n$, is it always possible to smoothly deform γ into $\tilde{\gamma}$ through intermediate curves?" For convenience, we assume that γ and $\tilde{\gamma}$ have the same parameter interval.

Definition 3.9

A **regular homotopy** between two curves $\gamma, \tilde{\gamma} : [a, b] \to \mathbb{R}^n$ is a one-parameter family $t \mapsto \gamma_t$ of curves $\gamma_t : [a, b] \to \mathbb{R}^n$, defined for $t \in [0, 1]$, such that $\gamma_0 = \gamma$ and $\gamma_1 = \tilde{\gamma}$. If there exists such a regular homotopy, γ and $\tilde{\gamma}$ are called **regularly homotopic**.

Regular homotopy is an equivalence relation on the set of curves $\gamma \colon [a, b] \to \mathbb{R}^n$: Reflexivity and symmetry are easy and for transitivity we make use (see Appendix A.2) of a smooth function $h \colon [0, 1] \to [0, 1]$ such that

$$h(x) = \begin{cases} 0, & \text{for } x \in [0, \epsilon] \\ 1, & \text{for } x \in [1 - \epsilon, 1]. \end{cases}$$

If now $t \mapsto \gamma_t$ is a regular homotopy between γ and $\hat{\gamma}$ and $t \mapsto \tilde{\gamma}_t$ a regular homotopy between $\hat{\gamma}$ and $\tilde{\gamma}$ then

$$t \mapsto \begin{cases} \gamma_{h(2t)}, & \text{for } t \in \left[0, \frac{1}{2}\right] \\ \tilde{\gamma}_{h(2t-1)}, & \text{for } t \in \left(\frac{1}{2}, 1\right] \end{cases}$$

is a regular homotopy from γ to $\tilde{\gamma}$. One can think of regular homotopies as smooth paths in the space of all curves $\gamma \colon [a, b] \to \mathbb{R}^n$, the equivalence classes under regular homotopy are the path-connected components of this space. Indeed, this space is connected, as we will prove below for the case $n = 2$. Using the curvature function for curves in $\mathbb{R}^n$ that will be introduced in Sect. 4.2, it would not be difficult to modify the proof and show that any two curves $\gamma \colon [a, b] \to \mathbb{R}^n$ are regularly homotopic.

Theorem 3.10
Any two curves $\gamma, \tilde{\gamma} \colon [a, b] \to \mathbb{R}^2$ are regularly homotopic.

Proof. By transitivity, we can construct the desired regular homotopy in steps. As a first step we use a regular homotopy to achieve that γ has length $b - a$:

$$\gamma_t = \left(1 - t + t\frac{b - a}{\mathcal{L}(\gamma)}\right)\gamma$$

Therefore, without loss of generality we may assume that the original curve γ already has length b-a. Then we can use a regular homotopy to achieve that γ has unit speed: using the inverse of the arclength function $s \colon [a, b] \to [0, L]$ of γ (Definition 1.13), we define a regular homotopy

$$\gamma_t(x) = \gamma((1 - t)x + t\, s^{-1}(x)).$$

So we can assume without loss of generality that γ has unit speed. Now we use a regular homotopy in order to translate the starting point of γ to the origin and achieve $\gamma(a) = 0$:

$$\gamma_t = (1 - t)\gamma(a) + \gamma.$$

Similarly, we can rotate γ to achieve that the unit tangent

$$T(a) = \begin{pmatrix} \cos \beta \\ \sin \beta \end{pmatrix}$$

of γ at the starting point becomes the first standard basis vector $\mathbf{e}_1$ of $\mathbb{R}^2$ (Fig. 3.7):

$$\gamma_t = \begin{pmatrix} \cos((1-t)\beta) & \sin((1-t)\beta) \\ -\sin((1-t)\beta) & \cos((1-t)\beta) \end{pmatrix} \gamma$$

We apply the same normalizations to $\tilde{\gamma}$. Now we consider the linear interpolation

$$\kappa_t = (1-t)\kappa + t\tilde{\kappa}$$

between the curvature functions κ of γ and $\tilde{\kappa}$ of $\tilde{\gamma}$ and define the desired regular homotopy from γ to $\tilde{\gamma}$ by

$$\alpha_t(x) := \int_0^x \kappa_t$$

$$T_t := \begin{pmatrix} \cos \alpha_t \\ \sin \alpha_t \end{pmatrix}$$

$$\gamma_t(x) := \int_0^x T_t.$$

$\square$

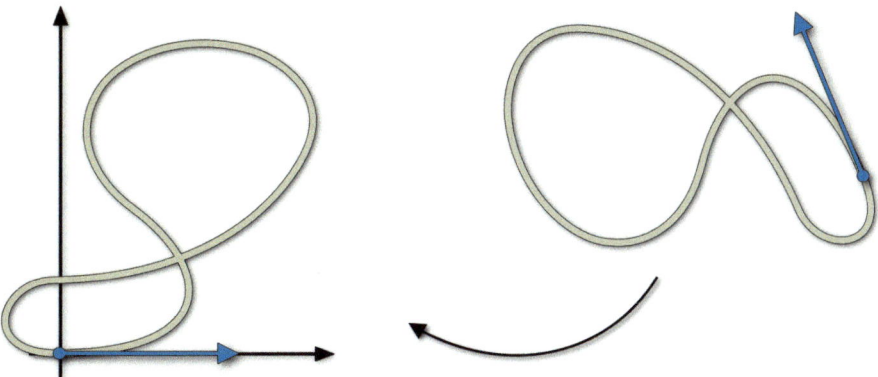

Fig. 3.7 The initial regular homotopy that brings the curve into a standard position and size

3.6   Whitney-Graustein Theorem

Definition 3.11

A **regular homotopy through closed curves** between two closed curves $\gamma, \tilde{\gamma} \colon [a, b] \to \mathbb{R}^2$ is a regular homotopy $t \mapsto \gamma_t$ between γ and $\tilde{\gamma}$ such that for all $t \in [0, 1]$ the curve γ_t is closed. If there exists such a regular homotopy, γ and $\tilde{\gamma}$ are called **regularly homotopic through closed curves**.

Let us start with an example that will be needed below. Recall that $\gamma \colon [a, b] \to \mathbb{R}^n$ is called closed if $\gamma = \hat{\gamma}|_{[a,b]}$ for some periodic smooth map $\hat{\gamma} \colon \mathbb{R} \to \mathbb{R}^n$ with period $b - a$. A simple way to make a new closed curve $\tilde{\gamma} \colon [a, b] :\to \mathbb{R}^n$ out of such a curve γ is by a so-called parameter shift, which depends on a number $\tau \in \mathbb{R}$:

$$\tilde{\gamma}(x) := \hat{\gamma}(x - \tau).$$

This closed curve $\tilde{\gamma}$ is regularly homotopic through closed curves to γ, a suitable regular homotopy being $t \mapsto \gamma_t$ with

$$\gamma_t(x) = \hat{\gamma}(x - t\tau).$$

Like regular homotopy in Sect. 3.5, regular homotopy as closed curves is an equivalence relation on the set of closed curves in $\mathbb{R}^2$ and the equivalence classes can be thought of as the connected components of this space. This time however, the whole space is not connected:

Theorem 3.12 ([45])
Two closed curves $\gamma, \tilde{\gamma} \colon [a, b] \to \mathbb{R}^2$ are regularly homotopic through closed curves if and only if they have the same tangent winding number.

In Fig. 3.8 we see an example of a regular homotopy through closed curves.

Proof. Suppose there is a regular homotopy as closed curves $t \mapsto \gamma_t$ between γ and $\tilde{\gamma}$. Denote by $ds_t = |\gamma_t'|$ and κ_t the speed and the curvature of γ_t. Then the tangent winding number

$$n_t = \frac{1}{2\pi} \int_a^b \kappa_t \, ds_t$$

is an integer for all $t \in [0, 1]$ and it depends continuously on t. Therefore it is constant and $n_0 = n_1$ means that γ and $\tilde{\gamma}$ they have the same tangent winding number.

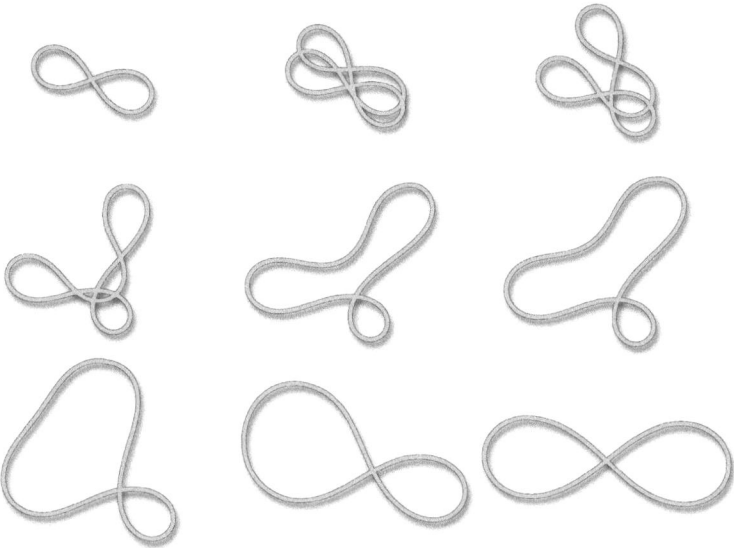

Fig. 3.8 A sequence of curves from a regular homotopy between the elastic figure eight curve traversed twice *(top left)* and the elastic figure eight curve traversed only once *(bottom right)*

Conversely, suppose that γ and $\tilde{\gamma}$ they have the same tangent winding number. As in the proof of Theorem 3.10 we can assume without loss of generality that γ and $\tilde{\gamma}$ both have unit speed. By Lemma 3.13 below and the fact that parameter shifts can be accomplished by regular homotopy as closed curves, we may also assume that the curvature functions κ and $\tilde{\kappa}$ are either constant or linearly independent. As in the proof of Theorem 3.10, we can apply another regular homotopy through closed curves to achieve $\gamma(a) = 0$ and $\gamma'(a) = \mathbf{e}_1$ (see Fig. 3.7). The same can be assumed for $\tilde{\gamma}$.

Let then $t \mapsto \gamma_t$ be the regular homotopy between γ and $\tilde{\gamma}$ constructed at the end of the proof of Theorem 3.10. The only problem is that for the intermediate curves γ_t might not be closed. We are going to repair this by modifying γ_t to a closed curve $\tilde{\gamma}_t$ as follows:

$$\tilde{\gamma}_t(x) = \gamma_t(x) - \frac{x-a}{b-a} \int_a^b T_t.$$

The only fact that needs to be checked is that $\tilde{\gamma}_t'(x) \neq 0$ for all $x \in [a, b]$.

Suppose we would have

$$0 = \tilde{\gamma}_t'(x) = T(x) - \frac{1}{b-a} \int_a^b T_t,$$

and therefore

$$1 = |T(x)| = \frac{1}{b-a}\left|\int_a^b T_t\right| \leq \frac{1}{b-a}\int_a^b |T_t| = 1.$$

The inequality sign in the above formula must be an equality, and this implies that T_t is constant, i.e.

$$0 = \kappa_t = (1-t)\kappa + t\tilde{\kappa}.$$

This would imply that κ and $\tilde{\kappa}$ are linearly dependent as functions, which by our assumptions means that κ and $\tilde{\kappa}$ are constant. Since both coefficients in the previous equation are positive, this would imply $\kappa = \tilde{\kappa} = 0$, which is impossible for closed curves. $\square$

As a consequence of Theorem 3.12, every closed curve in $\mathbb{R}^2$ is regularly homotopic through closed curves to one of the curves in the following list in Fig. 3.6.

We conclude this chapter with the Lemma that was needed in the proof of the Whitney-Graustein theorem:

Lemma 3.13
Let $\kappa\colon \mathbb{R} \to \mathbb{R}$ be a non-zero periodic function such that for all $\tau \in \mathbb{R}$ the functions κ and $x \mapsto \kappa(x - \tau)$ are linearly dependent. Then κ is constant.

Proof. Given our assumptions, there is a smooth function $\lambda\colon \mathbb{R} \to \mathbb{R}$ such that for all $x \in \mathbb{R}$ we have

$$\kappa(x - \tau) = \lambda(\tau)\kappa(x).$$

Differentiation with respect to τ at $\tau = 0$ yields

$$\kappa'(x) = \lambda'(0)\kappa(x)$$

The only non-zero periodic functions that satisfy such a differential equations are the constant functions. $\square$

Suppose we have a diffeomorphism $g\colon M \to \mathbb{R}^2$ where

$$M := \{x \in \mathbb{R}^2 \,|\, |x| \leq 1\}$$

is the unit disk in $\mathbb{R}^2$. Then we can define a closed curve $\gamma\colon [a, b] \to \mathbb{R}^2$ in such a way that the Fig. 3.9 becomes a commutative diagram.

Fig. 3.9 A diffeomorphism g from the unit disk M into $\mathbb{R}^2$ and the corresponding boundary loop γ

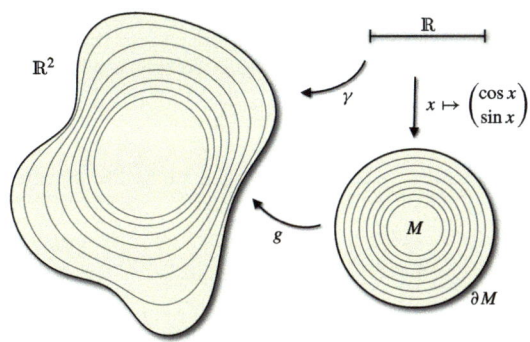

If a closed curve bounds a region in $\mathbb{R}^2$ that can be mapped onto the unit disk by a diffeomorphism, its tangent winding number is one or minus one:

Theorem 3.14

In the setup of Fig. 3.9, the tangent winding number of γ is ± 1, where the plus sign applies if and only if g preserves orientation, i.e. if $\det g'(x) > 0$ for all $x \in M$.

Proof. Already in the proof of Theorem 3.12 we saw that applying a scale or a rotation to γ does not change the regular homotopy class of γ. Therefore, without loss of generality we may assume

$$g'(0)\mathbf{e}_1 = \mathbf{e}_1.$$

For $t \in [0, 1]$ let A_t be the 2×2-matrix such that

$$A_t \mathbf{e}_1 = \mathbf{e}_1$$
$$A_t\, g'(\mathbf{e}_2) = (1 - t)g'(\mathbf{e}_2) \pm t\mathbf{e}_2$$

where the plus sign is chosen if and only if $\det g'(0) > 0$. Then the matrix A_t is invertible for all t and the one-parameter family $t \mapsto \gamma_t = A_t \gamma$ of closed curves is a regular homotopy, so after replacing g with $A_1 \circ g$ we can assume without loss of generality that

$$g'(0) = I.$$

Now define for $r \in (0, 1]$ closed curves $\gamma_r : [0, 2\pi] \to \mathbb{R}^2$ by

$$\gamma_r(x) = g\left(\begin{pmatrix} \cos(rx) \\ \sin(rx) \end{pmatrix}\right).$$

For small $\epsilon > 0$ the curve γ_ϵ is close the parametrization

$$x \mapsto \begin{pmatrix} \cos(x) \\ \pm \sin(x) \end{pmatrix}$$

of to the unit circle, so the tangent winding number of γ_ϵ is ± 1. On the other hand, $\gamma = \gamma_1$ is regularly homotopic to γ_ϵ, and therefore also the tangent winding number of γ is ± 1 (Fig. 3.8). $\qquad \square$

Parallel Normal Fields

4

For curves $\gamma \colon [a, b] \to \mathbb{R}^n$ there is an analog $\kappa \colon [a, b] \to \mathbb{R}^{n-1}$ of the curvature function of a plane curve. In the context of unit speed curves, this function κ determines γ up to an orientation-preserving rigid motion of $\mathbb{R}^n$. Before we can define κ, we have to study *parallel normal vector fields* along a curve in $\mathbb{R}^n$.

4.1   Parallel Transport

Definition 4.1

Let $\gamma \colon [a, b] \to \mathbb{R}^n$ be an immersion with unit tangent field $T \colon [a, b] \to \mathbb{R}^n$. Then a smooth map $Z \colon [a, b] \to \mathbb{R}^n$ is called a **normal field** for γ if

$$\langle Z(x), T(x) \rangle = 0$$

for all $x \in [a, b]$. The $(n-1)$-dimensional linear subspace $T(x)^\perp$ is called the **normal space** of γ at x.

If $Z \colon [a, b] \to \mathbb{R}^n$ is a normal field for γ, then we can split its derivative Z' into its tangential part and its normal part:

$$Z' = \lambda T + W$$

where $\lambda \colon [a, b] \to \mathbb{R}$ is a smooth function and W is another normal field. It turns out that $\lambda(x)$ can be computed from $Z(x)$ alone, without taking the derivative of Z: differentiating the expression $\langle Z, T \rangle = 0$ we obtain

$$\lambda = \langle Z', T \rangle = -\langle Z, T' \rangle.$$

© The Author(s) 2024
U. Pinkall, O. Gross, *Differential Geometry*, Compact Textbooks in Mathematics,
https://doi.org/10.1007/978-3-031-39838-4_4

The scalar product $\langle Z', Z \rangle = \frac{1}{2} \langle Z, Z \rangle'$ measures how the length of Z changes along γ. A component of Z' orthogonal to Z and T indicates a rotation of Z around the tangent T. If Z has constant length and there is no such twisting, Z is called parallel:

Definition 4.2

A **normal field** $Z \colon [a, b] \to \mathbb{R}^n$ along a curve $\gamma \colon [a, b] \to \mathbb{R}^n$ with unit tangent field $T \colon [a, b] \to \mathbb{R}^n$ is called **parallel** if there is a function $\lambda \colon [a, b] \to \mathbb{R}$ such that

$$Z' = \lambda T.$$

There is a parallel normal field Z for every immersion γ and all such fields come in an $(n - 1)$-parameter family:

Theorem 4.3
Given a vector $W \in T(a)^{\perp}$ in the normal space of a curve $\gamma \colon [a, b] \to \mathbb{R}^n$ at a, there is a unique parallel normal field $Z \colon [a, b] \to \mathbb{R}^n$ of γ such that

$$Z(a) = W.$$

If Z, Y are two parallel normal fields along γ, their scalar product $\langle Z, Y \rangle$ is constant.

Proof. If Z is a parallel normal vector field along γ with $Z(a) = W$, then differentiating the equation $\langle Z, T \rangle = 0$ yields $\langle Z', T \rangle + \langle Z, T' \rangle = 0$ and, using $Z' = -\langle Z, T' \rangle T$, we see that Z solves the linear initial value problem

$$Z(a) = W$$
$$Z' = \langle Z, T \rangle T' - \langle Z, T' \rangle T.$$

By the Picard-Lindelöf theorem, such a solution is unique, which proves the uniqueness part of our claim. For the existence part, let Z be the solution of the above initial value problem. For any further solution Y of the above differential equation we have

$$\begin{aligned}
\langle Z, Y \rangle' &= \langle Z', Y \rangle + \langle Z, Y' \rangle \\
&= \langle \langle Z, T \rangle T' - \langle Z, T' \rangle T, Y \rangle + \langle Z, \langle Y, T \rangle T' - \langle Y, T' \rangle T \rangle \\
&= 0.
\end{aligned}$$

and therefore the scalar product $\langle Z, Y \rangle$ is constant. In particular, $Y = T$ is such a solution, so $\langle Z(a), T(a) \rangle = 0$ implies $\langle Z, T \rangle = 0$. Therefore Z is a normal field, in fact a parallel one. $\qquad\qquad\qquad\qquad\qquad\qquad\qquad\qquad\qquad\qquad\qquad\qquad\qquad\qquad\qquad\square$

If $\gamma : [a, b] \to \mathbb{R}^n$ is a curve and W is a vector in $T(a)^\perp$, for every $x \in [a, b]$ we can use the parallel normal field Z with $Z(a) = W$ to "transport" W to a normal vector $Z(x) \in T(x)^\perp$. This parallel transport map

$$P(x) \colon T(a)^\perp \to T(x)^\perp$$

is obviously linear, and by Theorem 4.3 it is in fact orthogonal, i.e. it preserves scalar products. Moreover, each normal space $T(x)^\perp$ carries an orientation in the sense that a basis $W_1, \ldots, W_{n-1}$ of $T(x)^\perp$ is called positively oriented if

$$\det(W_1, \ldots, W_{n-1}, T(x)) > 0.$$

If $W_1, \ldots, W_{n-1}$ is a positively oriented basis of $T(a)^\perp$ and $Z_1, \ldots, Z_{n-1}$ are parallel normal fields with $Z_j(a) = Y_j$ then the map

$$x \mapsto \det(Z_1(x), \ldots, Z_{n-1}(x), T(x))$$

is continuous and never zero. Therefore, for all $x \in [a, b]$ we have

$$\det(Z_1(x), \ldots, Z_{n-1}(x), T(x)) > 0$$

and the map $P(x)$ is orientation-preserving. We summarize this as follows:

Definition 4.4

Given a curve $\gamma : [a, b] \to \mathbb{R}^n$ and $x \in [a, b]$, the orientation-preserving orthogonal map $P(x) \colon T(a)^\perp \to T(x)^\perp$ defined above is called the **parallel transport** from the normal space $T(a)^\perp$ to the normal space $T(x)^\perp$.

By Theorem 4.3, each vector $Z(x)$ of a parallel normal field has the same length. Therefore, we can use parallel normal fields Z in order to displace a curve γ by a fixed distance $\epsilon = |Z|$, without introducing unnecessary twisting:

Definition 4.5

If Z is a parallel normal field along a curve $\gamma : [a, b] \to \mathbb{R}^n$ and the derivative of

$$\tilde{\gamma} = \gamma + Z$$

vanishes nowhere, then the $\tilde{\gamma}$ is called a **parallel curve** of γ.

For a curve $\gamma : [a, b] \to \mathbb{R}^n$ the continuous (but not necessarily smooth) function

$$\left| \frac{dT}{ds} \right| : [a, b] \to \mathbb{R}$$

is called the **absolute curvature** of γ. If $\epsilon > 0$ is such that

$$\frac{1}{\epsilon} > \max \left\{ \left| \frac{dT}{ds}(x) \right| \ \Big| \ x \in [a, b] \right\}$$

and Z is a parallel normal field with $|Z| = \epsilon$ then by the Cauchy-Schwarz inequality we have

$$\frac{d(\gamma + Z)}{ds} = \left(1 + \left\langle Z, \frac{dT}{ds} \right\rangle \right) T \neq 0.$$

Therefore, if we pick a vector $W \in T(a)^\perp$ with sufficiently small norm and define Z as the parallel normal field Z with $Z(a) = W$, then $\gamma + Z$ will be a parallel curve for γ.

As an application, we always visualize a curve in $\mathbb{R}^3$ by thickening it, which means that we chose a suitable collection of $W \in T(a)^\perp$ with small length and draw the union of the corresponding parallel normal fields. Most of the time we use a small circle centered at the origin in $T(a)^\perp$, but different choices (as in Fig. 4.1) are also possible.

Fig. 4.1 A closed curve in the normal space $T(a)^\perp$ is used to build a thickened version of γ by parallel transport

4.2   Curvature Function of a Curve in $\mathbb{R}^n$

We saw in Sect. 3.1 that, up to rigid motions of $\mathbb{R}^2$, the geometry of a unit speed curve $\gamma : [a, b] \rightarrow \mathbb{R}^2$ is completely determined by its curvature function $\kappa : [a, b] \rightarrow \mathbb{R}$. Here we will define a similar curvature function $\kappa : [a, b] \rightarrow \mathbb{R}^{n-1}$ for any unit speed curve $\gamma : [a, b] \rightarrow \mathbb{R}^n$. To define $\kappa(x)$, we use parallel transport to transfer the normal vector $T'(x) \in T(x)^\perp$ to the normal space $T(a)^\perp$. Afterwards we use an orthonormal basis of $T(a)^\perp$ in order to identify $T(a)^\perp$ with $\mathbb{R}^{n-1}$.

> **Theorem 4.6**
> Let $\gamma : [a, b] \rightarrow \mathbb{R}^n$ be a curve with unit tangent T and parallel transport maps $P(x) : T(a)^\perp \rightarrow T(x)^\perp$. Then there is a unique smooth map $\Psi : [a, b] \rightarrow T(a)^\perp$ such that for all $x \in [a, b]$ we have
>
> $$P(x)(\Psi(x)) = -\frac{dT}{ds}(x).$$
>
> Ψ is called the **Hasimoto curvature** of γ.

See Sect. 5.3 for the details on Hasimoto's contribution. The Hasimoto curvature determines γ uniquely:

> **Theorem 4.7**
> Given a point $\mathbf{p} \in \mathbb{R}^n$, a unit vector $S \in \mathbb{R}^n$ and a smooth map $\Psi : [a, b] \rightarrow T(a)^\perp$, there is a unique unit speed curve $\gamma : [a, b] \rightarrow \mathbb{R}^n$ such that $\gamma(a) = \mathbf{p}$, $\gamma'(a) = S$ and Ψ is the Hasimoto curvature of γ (see Fig. 4.2).

Fig. 4.2 The Hasimoto curvature Ψ of a curve γ indicated as a blue curve in $T(a)^\perp$

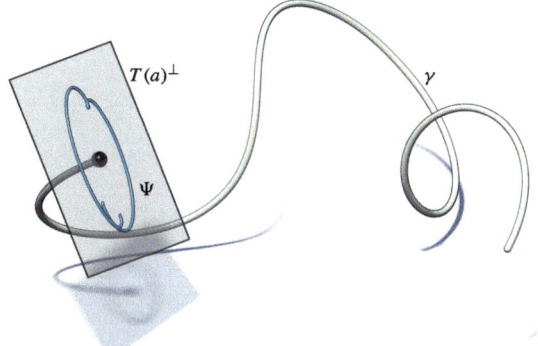

Proof. First we prove uniquess of γ. Let $\gamma : [a, b] \to \mathbb{R}^n$ be a curve with the desired properties. Choose an orthonormal basis $W_1, \ldots, W_{n-1}$ of $T(a)^{\perp}$ such that

$$\det(W_1, \ldots, W_{n-1}, T(a)) = 1$$

and define $\kappa_1, \ldots \kappa_{n-1}$ by

$$\Psi = \kappa_1 W_1 + \ldots + \kappa_{n-1} W_{n-1}.$$

Let $Z_1, \ldots, Z_{n-1}$ be the parallel normal fields along γ such that $Z_j(a) = W_j$ for all $j \in \{1, \ldots, n-1\}$. Then

$$(Z_1, \ldots, Z_{n-1}, T) : [a, b] \to \mathbb{R}^{n \times n}$$

solves the initial value problem

$$(Z_1, \ldots, Z_{n-1}, T)(a) = (W_1, \ldots, W_{n-1}, S)$$

$$(Z_1, \ldots, Z_{n-1}, T)' = \left(\kappa_1 T, \ldots, \kappa_{n-1} T, -\sum_{j=1}^{n-1} \kappa_j Z_j \right)$$

and is therefore uniquely determined by $\mathbf{p}$, S and Ψ. In particular, T is uniquely determined and so is

$$x \mapsto \gamma(x) = \int_a^x T.$$

For existence, we can use the above initial value problem to define the vector fields $(Z_1, \ldots, Z_{n-1}, T)$. At $x = a$ these vectors are orthonormal and their pairwise scalar products solve the system of linear differential equations

$$\langle T, T \rangle' = -2 \sum_{j=1}^{n-1} \kappa_j \langle T, Z_j \rangle$$

$$\langle T, Z_j \rangle' = \kappa_j \langle T, T \rangle - \sum_{i=1}^{n-1} \kappa_i \langle Z_i, Z_j \rangle$$

$$\langle Z_i, Z_j \rangle' = \kappa_i \langle T, Z_j \rangle + \kappa_j \langle Z_i, T \rangle.$$

We can interpret this as an initial value problem for the functions $\langle T, Z_j \rangle$, $\langle T, T \rangle$ and $\langle Z_i, Z_j \rangle$. The functions $\langle T, Z_j \rangle = 0$, $\langle T, T \rangle = 1$, $\langle Z_i, Z_j \rangle = \delta_{ij}$ solve this initial value problem, and by Picard and Lindelöf such a solution is unique. Therefore, $(Z_1, \ldots, Z_{n-1}, T)$ stay orthonormal. So by integration of T we obtain a unit speed curve $\gamma : [a, b] \to \mathbb{R}^n$ with $\gamma(a) = \mathbf{p}$ and $\gamma'(a) = S$. $Z_1, \ldots Z_{n-1}$ are parallel normal fields along γ with $Z_j(a) = W_j$. Because we already know that $T' = -\sum_{j=1}^{n-1} \kappa_j Z_j$, this implies that Ψ is indeed the Hasimoto curvature of γ.   □

In the above proof we used a basis of $T(a)^\perp$ in order to turn Ψ into an $\mathbb{R}^{n-1}$-valued function κ. This function is the promised analog of the curvature function of a plane curve:

Definition 4.8

Let $\gamma : [a, b] \to \mathbb{R}^n$ be a unit speed curve with unit tangent T and Hasimoto curvature Ψ. Let $W_1, \ldots, W_{n-1}$ be a positively oriented orthonormal basis of $T(a)^\perp$. Then the function

$$\kappa : [a, b] \to \mathbb{R}^{n-1}, \quad \kappa = \begin{pmatrix} \kappa_1 \\ \vdots \\ \kappa_{n-1} \end{pmatrix}$$

defined by

$$\Psi = \kappa_1 W_1 + \ldots + \kappa_{n-1} W_{n-1}$$

is called a **curvature function** of γ.

In the case $n = 2$ the positively oriented orthonormal basis of $T(a)^\perp$ mentioned in the above definition is unique, and therefore each plane curve has a unique curvature function $\kappa : [a, b] \to \mathbb{R}^1 = \mathbb{R}$, which is the one we already encountered in Sect. 3.1. It is clear from its definition that for any n the function κ is at least unique up to a rotation of $\mathbb{R}^{n-1}$:

Theorem 4.9
If $\kappa, \tilde{\kappa} : [a, b] \to \mathbb{R}^{n-1}$ are curvature functions of the same curve $\gamma : [a, b] \to \mathbb{R}^n$, then there is an orthogonal $((n-1) \times (n-1))$-matrix A with $\det A = 1$ such that

$$\tilde{\kappa} = A\kappa.$$

On the other hand, as in the case of curves in $\mathbb{R}^2$, for every curvature function $\kappa : [a, b] \to \mathbb{R}^{n-1}$ there is a corresponding curve $\gamma : [a, b] \to \mathbb{R}^n$ and γ is unique up to post-composition with an orientation preserving rigid motion of $\mathbb{R}^n$. Also the following theorem is a direct consequence of Theorem 4.7:

Theorem 4.10

Given a smooth function $\kappa\colon [a, b] \to \mathbb{R}^{n-1}$, there is a unit speed curve $\gamma\colon [a, b] \to \mathbb{R}^n$ for which κ is a curvature function. The curve γ is unique up to an orientation preserving rigid motion of $\mathbb{R}^n$, which means that if $\tilde{\gamma}$ is another curve having κ as a curvature function, then there is an orthogonal $(n \times n)$-matrix A with $\det A = 1$ and a vector $\mathbf{b} \in \mathbb{R}^n$ such that

$$\tilde{\gamma} = A\gamma + \mathbf{b}.$$

4.3   Geometry in Terms of the Curvature Function

Let $\gamma\colon [a, b] \to \mathbb{R}^n$ be a unit speed curve with unit tangent field T and $W_1, \ldots, W_{n-1}$ a positively oriented orthonormal basis of $T(a)^\perp$. Let $Z_1, \ldots, Z_{n-1}$ be the corresponding parallel normal fields along γ with $Z_j(a) = W_j$. Then we can describe every normal field Y along γ in terms of a function $y\colon [a, b] \to \mathbb{R}^n$ as

$$Y = y_1 Z_1 + \ldots y_{n-1} Z_{n-1}$$

$$= \begin{pmatrix} | & & | \\ Z_1 & \ldots & Z_{n-1} \\ | & & | \end{pmatrix} \begin{pmatrix} y_1 \\ \vdots \\ y_{n-1} \end{pmatrix}$$

$$=: Ny$$

where for $x \in [a, b]$ the matrix $N(x)$ has the vectors $Z_1(x), \ldots, Z_{n-1}(x) \in \mathbb{R}^n$ as its column vectors. In terms of the curvature function κ introduced in Definition 4.8 the derivative of Y can be expressed as

$$Y' = \langle \kappa, y \rangle T + Ny'.$$

In particular, for $Y = T'$ we obtain

$$T' = -N\kappa$$

$$T'' = -\langle \kappa, \kappa \rangle T - N\kappa'$$

$$T''' = -3\langle \kappa, \kappa' \rangle T + N\left(\langle \kappa, \kappa \rangle \kappa - \kappa''\right).$$

Now we are able to generalize the results we obtained in Sect. 3.3 for plane curves:

Theorem 4.11
A unit speed curve $\gamma : [a, b] \to \mathbb{R}^n$ is torsion-free elastic if and only if there is a constant $\lambda \in \mathbb{R}$ such that its curvature function κ satisfies

$$\kappa'' + \frac{|\kappa|^2}{2}\kappa + \lambda\kappa = 0.$$

Proof. By Theorem 2.23, γ is torsion-free elastic if and only if there is a constant $\lambda \in \mathbb{R}$ such that

$$0 = T''' + 3\langle T', T''\rangle T + \frac{3}{2}\langle T', T'\rangle T' - \lambda T'$$

$$= -N\left(\kappa'' + \frac{|\kappa|^2}{2}\kappa + \lambda\kappa\right).$$

$\square$

Here are further examples of how the geometry of γ is reflected in the properties of κ:

Theorem 4.12
Let $\kappa : [a, b] \to \mathbb{R}^{n-1}$ be a curvature function of a unit speed curve $\gamma : [a, b] \to \mathbb{R}^n$. Then:

   (i) $\kappa = 0$ if and only if the image of γ lies on a straight line.
   (ii) κ is a non-zero constant if and only if the image of γ lies on a circle.
   (iii) The image of κ lies in a hyperplane through the origin of $\mathbb{R}^{n-1}$ if and only if the image of γ lies in a hyperplane of $\mathbb{R}^n$.
   (iv) The image of κ lies in a hyperplane of $\mathbb{R}^{n-1}$ that does not pass through the origin if and only if the image of γ lies in a hypersphere of $\mathbb{R}^n$ (see Fig. 4.3).

Fig. 4.3 The curvature
function κ of a curve on S^2
lies on a straight line which
does not pass through the
origin

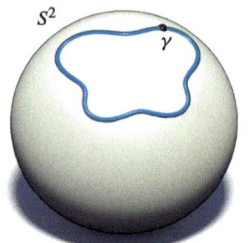 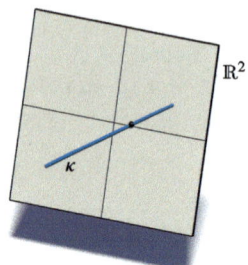

Proof. Claim (i) is obvious, since the image of a curve lies on a straight line if and
only if its unit tangent T is constant. If the image of κ lies in a hyperplane through
the origin of $\mathbb{R}^{n-1}$, there is a unit vector $\mathbf{a} \in \mathbb{R}^{n-1}$ such that $\langle \mathbf{a}, \kappa \rangle = 0$. Then

$$(N\mathbf{a})' = -\langle \kappa, \mathbf{a} \rangle = 0$$

so there is a fixed vector $\mathbf{n} \in \mathbb{R}^n$ such that $N\mathbf{a} = \mathbf{n}$. We have

$$\langle \mathbf{n}, \gamma \rangle' = \langle N\mathbf{a}, T \rangle = 0$$

and therefore the image of γ is contained in a hyperplane with normal vector $\mathbf{n}$. The
proof of the converse is left to the reader. This establishes (iii). For (iv), suppose that
there is a unit vector $\mathbf{a} \in \mathbb{R}^{n-1}$ and a number $r > 0$ such that

$$\langle \mathbf{a}, \kappa \rangle = \frac{1}{r}.$$

Then

$$(\gamma - rN\mathbf{a})' = T - r\langle \kappa, \mathbf{a} \rangle T = 0$$

so there is a fixed point $\mathbf{m} \in \mathbb{R}^n$ such that

$$\gamma - rN\mathbf{a} = \mathbf{m}$$

and we have

$$|\gamma - \mathbf{m}| = r.$$

Therefore, the image of γ lies on the hypersphere with center $\mathbf{m}$ and radius r. Again,
the proof of the converse is left to the reader and we have established (iv). For (ii)
we use induction on n based on (iii), starting at $n = 2$ where we use (iv). $\square$

Curves in $\mathbb{R}^3$

5

For a closed curve $\gamma \colon [a, b] \to \mathbb{R}^3$ with unit tangent field T, we can use a parallel normal vector field to transport a normal vector $W_a \in T(a)^\perp$ at the starting point of γ to a normal vector $W_b \in T(b)^\perp$ at the end point. If γ is closed, the angle $\mathcal{T}(\gamma)$ between W_b and W_a is called the *total torsion* of γ. A notion of total torsion can also be defined for curves in $\mathbb{R}^3$ that are not necessarily closed. Therefore, for curves in $\mathbb{R}^3$, total torsion provides another geometric functional besides length or bending energy. Critical points of a linear combination of length, total torsion and bending energy are needed for modelling the physical equilibrium shapes of elastic wires in $\mathbb{R}^3$.

5.1 Total Torsion of Curves in $\mathbb{R}^3$

Let us focus now on curves $\gamma \colon [a, b] \to \mathbb{R}^3$. In this case we can visualize the parallel transport of normal directions introduced in Sect. 4.1 as approximately implemented by a chain of so-called "constant velocity joints" (cf. [35]). Such joints are able to transport normal directions in an angle-preserving manner. Rotating the initial vector $Z(a)$ of a parallel normal field by an angle α will make the final vector $Z(b)$ rotate by the same angle α (see Figs. 5.1 and 5.2).

Definition 5.1

For a curve $\gamma \colon [a, b] \to \mathbb{R}^3$ the orthogonal linear map

$$\mathcal{P} \colon T(a)^\perp \to T(b)^\perp, \ X \mapsto Z(b)$$

where $Z \colon [a, b] \to \mathbb{R}^3$ is the parallel normal field along γ with $Z(a) = X$ is called the **normal transport** of γ.

U. Pinkall, O. Gross, *Differential Geometry*, Compact Textbooks in Mathematics,
https://doi.org/10.1007/978-3-031-39838-4_5

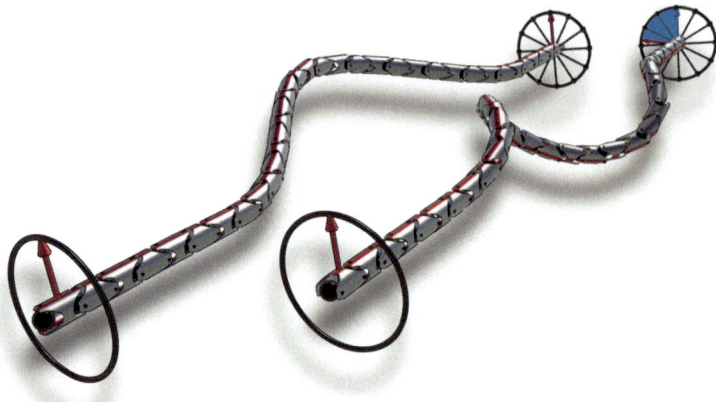

Fig. 5.1 Two curves built from constant-velocity joints. The total torsion is zero for the curve on the left, but not for the one on the right. The red line indicates a parallel normal field

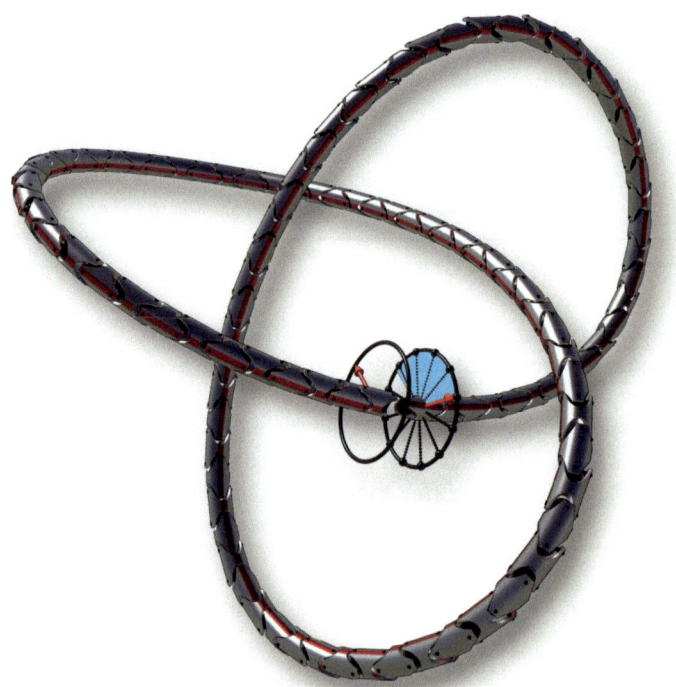

Fig. 5.2 A trefoil knot cannot be built from constant-velocity joints. Because of the angle-defect due to the total torsion, the joints would not close up

After having chosen a pair $W = (W_a, W_b)$ of unit vectors $W_a \in T(a)^\perp$ and $W_b \in T(b)^\perp$ we can describe the normal transport $\mathcal{P}$ by an angle:

Definition 5.2

Let $\gamma : [a, b] \to \mathbb{R}^3$ be curve with unit tangent T and normal transport $\mathcal{P}$. Then, given a pair $W = (W_a, W_b)$ of unit vectors $W_a \in T(a)^\perp$ and W_b in $T(b)^\perp$, the unique angle

$$\mathcal{T}_W \in \mathbb{R}/2\pi\mathbb{Z}$$

with

$$\mathcal{P}(W_a) = (\cos \mathcal{T}_W)\ W_b + (\sin \mathcal{T}_W)\ T(b) \times W_b.$$

is called the **total torsion** (cf. Sect. 5.4) of the curve γ with respect to W_a and W_b.

For a closed curve we have $T(a) = T(b)$ and we can always choose $W(b) = W(a)$. The total torsion $\mathcal{T}_W$ then becomes independent of the choice of W_a, so in this case we can drop the subscript W and denote the total torsion of a closed curve γ by $\mathcal{T}(\gamma)$. Let us determine the infinitesimal variation of the total torsion $\mathcal{T}_W$ if we vary the curve γ as well as the unit vectors W_a and W_b:

Theorem 5.3

Let $t \mapsto \gamma_t$ be a variation with variational vector field $\dot{\gamma} = Y$ of a curve $\gamma : [a, b] \to \mathbb{R}^3$. Let $t \mapsto W_a(t) \in T_t(a)^\perp$ and $t \mapsto W_b(t) \in T_t(b)^\perp$ be two smooth families of unit vectors. Then the total torsion $\mathcal{T}_W(t)$ of γ_t with respect to

$$W(t) = (W_a(t), W_b(t))$$

satisfies

$$\frac{d}{dt}\bigg|_{t=0} \mathcal{T}_W(t) = \langle \dot{W}_a, T(a) \times W_a \rangle - \langle \dot{W}_b, T(b) \times W_b \rangle - \int_a^b \det(T, T', \dot{T}).$$

In terms of Y, the above integral can be expressed as

$$\int_a^b \det(T, T', \dot{T}) = \left\langle Y, T \times \frac{dT}{ds} \right\rangle\bigg|_a^b - \int_a^b \left\langle Y, T \times \left(\frac{dT}{ds}\right)' \right\rangle.$$

Proof. Let Z_t be the parallel normal field along γ_t with $Z_t(a) = W_a(t)$. In particular, $Z_0 =: Z$ is a parallel normal field and we have

$$Z' = -\langle Z, T' \rangle T.$$

Taking the time derivative of the equation

$$Z_t(b) = \cos(\mathcal{T}_W(t))\, W_b(t) + \sin(\mathcal{T}_W(t))\, T(b) \times W_b(t)$$

at $t = 0$ yields

$$\langle \dot{Z}(b), T(b) \times Z(b) \rangle = \dot{\mathcal{T}}_W + \langle \dot{W}_b, T(b) \times Z(b) \rangle.$$

From Theorem 5.4

$$\int_a^b \det(T, T', \dot{T}) = -\int_a^b \langle \dot{Z}, T \times Z \rangle'$$

$$= -\langle \dot{Z}(b), T(b) \times Z(b) \rangle + \langle \dot{Z}(a), T(a) \times Z(a) \rangle$$

$$= -\dot{\mathcal{T}}_W - \langle \dot{W}_b, T(b) \times Z(b) \rangle + \langle \dot{W}_a, T(a) \times Z(a) \rangle.$$

The second claim is a consequence of

$$\det(T, T', \dot{T}) = \det\left(T, T', \frac{dY}{ds}\right)$$

$$= \det\left(T, \frac{dT}{ds}, Y'\right)$$

$$= \det\left(T, \frac{dT}{ds}, Y\right)' - \det\left(T, \left(\frac{dT}{ds}\right)', Y\right),$$

where we used that $\dot{T} = \frac{dY}{ds} - \langle \frac{dY}{ds}, T \rangle T$ (cf. proof of Theorem 2.8). □

The following Theorem was used in the above proof and will be needed also in Sect. 5.3:

Theorem 5.4

Let $t \mapsto \gamma_t$ be a variation of a curve $\gamma : [a, b] \to \mathbb{R}^3$ with unit tangent field T. Let $t \mapsto Z_t$ be a smooth one-parameter family of maps such that Z_t is a parallel unit normal field along γ_t. Then

$$\langle \dot{Z}, T \times Z \rangle' = \langle T, \dot{T} \times T' \rangle.$$

Proof. The prime derivative commutes with the dot derivative, hence

$$\langle \dot{Z}, T \times Z \rangle' = \langle (Z')^{\boldsymbol{\cdot}}, T \times Z \rangle + \langle \dot{Z}, T' \times Z \rangle$$

$$= \Big(-\langle Z, T' \rangle \dot{T}, T \times Z \Big) + \langle \dot{Z}, T' \times Z \rangle.$$

Because $\langle Z_t, Z_t \rangle = 1$ and $\langle Z_t, T_t \rangle = 0$, for all t, we have

$$\dot{Z} = \langle \dot{Z}, T \rangle T + \langle \dot{Z}, T \times Z \rangle T \times Z$$

$$= -\langle Z, \dot{T} \rangle T + \langle \dot{Z}, T \times Z \rangle T \times Z$$

and therefore we can continue the previous calculation of $\langle \dot{Z}, T \times Z \rangle'$ as follows:

$$\langle \dot{Z}, T \times Z \rangle' = \langle Z, \dot{T} \rangle \langle T \times Z, T' \rangle - \langle Z, T' \rangle \langle T \times Z, \dot{T} \rangle$$

$$= \langle Z \times (T \times Z), \dot{T} \times T' \rangle$$

$$= \langle T, \dot{T} \times T' \rangle.$$

$\square$

In Sect. 4.3 we described normal vector fields Y along a curve $\gamma : [a, b] \rightarrow \mathbb{R}^n$ in terms of functions $y : [a, b] \rightarrow \mathbb{R}^{n-1}$. Given a parallel unit normal field Z along a curve $\gamma : [a, b] \rightarrow \mathbb{R}^3$ with unit tangent T, in terms of this correspondence, any normal field Y can be written as

$$Y = Ny = y_1 Z + y_2 T \times Z.$$

Then the function $\langle \dot{Z}, T \times Z \rangle$ in Theorem 5.3 also appears if, given a variation of γ and Z, we want to know the time derivative of the above formula:

Theorem 5.5

Let $t \mapsto \gamma_t$ be a variation of a curve $\gamma : [a, b] \rightarrow \mathbb{R}^3$ with unit tangent field T. Let $t \mapsto Z_t$ be a smooth one-parameter family of maps such that Z_t is a parallel unit normal field along γ_t and $t \mapsto y_t$ a smooth family of maps $y_t : [a, b] \rightarrow \mathbb{R}^2$. Then

$$(Ny)^{\boldsymbol{\cdot}} = -\langle Ny, \dot{T} \rangle T + N \left(\dot{y} + \langle \dot{Z}, T \times Z \rangle Jy \right).$$

Proof. Taking into account the time derivatives of the equations that tell us $T, Z, T \times Z$ are orthonormal, we obtain

$$(Ny)^{\bullet} = N\dot{y} + y_1(\langle \dot{Z}, T \rangle T + \langle \dot{Z}, T \times Z \rangle T \times Z)$$
$$+ y_2(\langle (T \times Z)^{\bullet}, T \rangle T + \langle (T \times Z)^{\bullet}, Z \rangle Z)$$
$$= -\langle Ny, \dot{T} \rangle T + N \left(\dot{y} + \langle \dot{Z}, T \times Z \rangle Jy \right).$$

$\square$

5.2   Elastic Curves in $\mathbb{R}^3$

The torsion-free elastic curves studied in the Sects. 2.4 and 2.5 were critical points of bending energy under the constraint of fixed length. For general elastic curves in $\mathbb{R}^3$ also the total torsion is constrained (see Fig. 5.3). Note that for a variation $t \mapsto \gamma_t$ with support in the interior of $[a, b]$ of a curve $\gamma : [a, b] \to \mathbb{R}^3$ with unit tangent T all parallel transport maps $\mathcal{P}_t$ are defined on the same vector space $T(a)^{\perp}$. Therefore it makes sense to consider the derivative $\frac{d}{dt}\big|_{t=0} \mathcal{P}_t$.

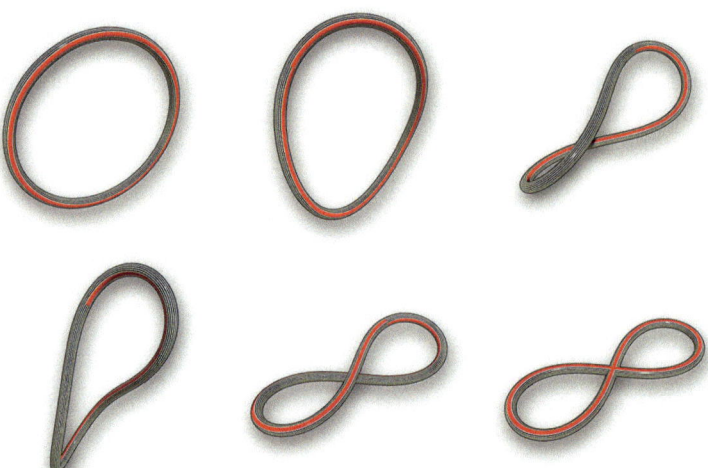

Fig. 5.3 Elastic curves obtained by minimizing bending energy under the constraint of fixed length and fixed total torsion, for various values of the total torsion constraint: $0, \frac{2}{5}\pi, \frac{6}{5}\pi, \frac{14}{10}\pi, \frac{9}{5}\pi, 2\pi$

Definition 5.6

A curve $\gamma : [a, b] \to \mathbb{R}^3$ is called an **elastic curve** if it is a critical point of the bending energy $\mathcal{B}$ under the constraint of fixed length $\mathcal{L}$ and fixed normal transport $\mathcal{P}$.

During a variation $t \mapsto \gamma_t$ of a curve $\gamma : [a, b] \to \mathbb{R}^3$ with constant support in the interior of $[a, b]$, after choosing unit vectors $W_a \in T(a)^\perp$ and $W_b \in T(b)^\perp$ (independent of t), we can measure the normal transport along γ_t as the total torsion angle $\mathcal{T}_W(\gamma_t)$. Theorem 5.3 then will tell us the infinitesimal variation of the normal transport, in a way that does not depend on the choice of W_a and W_b.

Theorem 5.7
Let $\gamma : [a, b] \to \mathbb{R}^3$ be a unit speed curve with unit tangent T. Then the following are equivalent:

(i) γ is an elastic curve.
(ii) There are constants $\lambda, \mu \in \mathbb{R}$ such that

$$T''' - \langle T''', T \rangle T + \frac{3}{2} \langle T', T' \rangle T' - \mu T \times T'' - \lambda T' = 0.$$

(iii) There are constants $\lambda, \mu \in \mathbb{R}$ and a constant vector $\mathbf{a} \in \mathbb{R}^3$ such that

$$T'' + \frac{3}{2} \langle T', T' \rangle T - \mu T \times T' - \lambda T + \mathbf{a} = 0.$$

(iv) There is a constant $\mu \in \mathbb{R}$ and a constant vector $\mathbf{a} \in \mathbb{R}^3$ such that

$$T'' - \langle T'', T \rangle T + \mathbf{a} - \langle \mathbf{a}, T \rangle T - \mu T \times T' = 0.$$

(v) There is a constant $\mu \in \mathbb{R}$ and constant vectors $\mathbf{a}, \mathbf{b} \in \mathbb{R}^3$ such that

$$\gamma' \times \gamma'' = \mu \gamma' + \mathbf{a} \times \gamma + \mathbf{b}.$$

(vi) There are constant vectors $\mathbf{a}, \mathbf{b} \in \mathbb{R}^3$ such that

$$\gamma'' = (\mathbf{a} \times \gamma + \mathbf{b}) \times \gamma'.$$

Proof. Theorem 2.21 was formulated in such a way that it is capable of dealing with several constraints, so that it is possible to prove Theorem 2.20 with more than just a single constraint. So more constraints than just the length are possible in Theorem 2.23. Therefore, the equivalence of (i) and (ii) can be shown following the same arguments that lead to Theorem 2.23. Here, taking derivatives of the equation $\langle T, T \rangle = 1$ gives

$$\langle T''', T \rangle = -3\langle T, T'' \rangle.$$

The equivalence of (ii) and (iii) follows from the equality

$$\left(T'' + \frac{3}{2}\langle T', T' \rangle T - \mu T \times T' - \lambda T \right)'$$

$$= T''' - \langle T''', T \rangle T + \frac{3}{2}\langle T', T' \rangle T' - \mu T \times T'' - \lambda T'$$

which can again be verified by taking derivatives of the equation $\langle T, T \rangle = 1$. (iv) is just the component of (iii) orthogonal to T, so it follows from (iii). In order to show that (iv) implies (iii), we have to show that (iv) implies that there is a constant λ such that also the equation obtained by taking the component of (iii) parallel to T is satisfied if (iv) holds, which is indeed the case:

$$\left(\langle T'', T \rangle + \frac{3}{2}\langle T', T' \rangle + \langle \mathbf{a}, T \rangle \right)' = \left(\frac{1}{2}\langle T', T' \rangle + \langle \mathbf{a}, T \rangle \right)'$$

$$= \langle T'' + a, T' \rangle$$

$$= \langle \mu T \times T', T' \rangle$$

$$= 0.$$

To prove that (iv) is equivalent to (v), note first that, the left-hand side of (iv) being orthogonal to T, (v) is equivalent to the equation obtained from (iv) by taking the cross product with T:

$$T \times T'' + T \times \mathbf{a} + \mu T' = 0$$

or

$$0 = (T \times T' + \gamma \times \mathbf{a} + \mu T)'.$$

Therefore, (iv) is equivalent to (v). Taking the cross product of the equation in (v) with T yields

$$-\gamma'' = -(\mathbf{a} \times \gamma) \times \gamma' = \mathbf{b} \times \gamma'$$

which is equivalent to (vi). To show that (vi) implies (v), note that the component orthogonal to T of the equation in (v) is equivalent to (vi). This means that we have

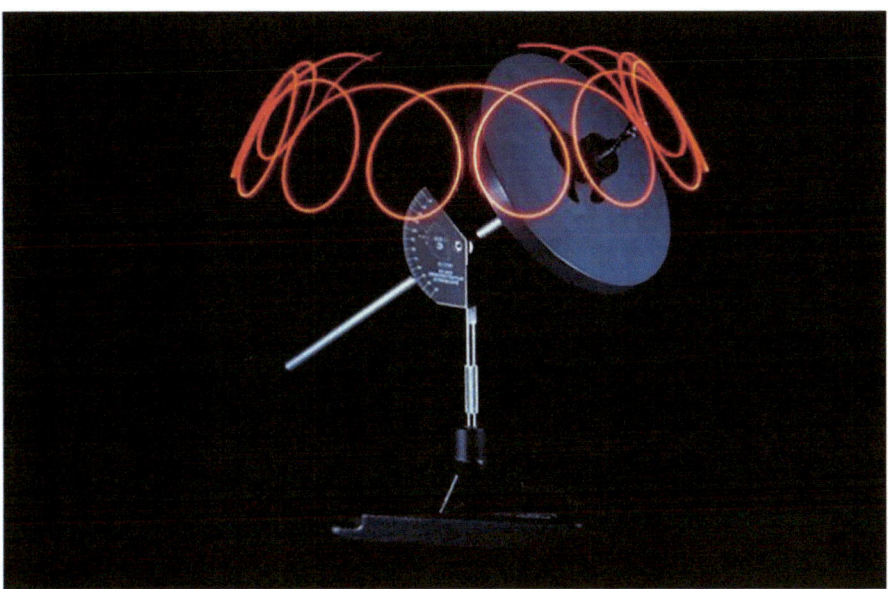

Fig. 5.4 Kirchhoff showed that, as an elastic curve is traversed with unit speed, its tangent vector T follows the motion of the axis of a gyroscope. The photograph with long-time exposure was reproduced from [41] with permission from Javier Villegas

only to show based on (vi) that the scalar product with T of the sum of the terms without the μT term is constant. This is indeed the case:

$$-\langle T, \gamma \times \mathbf{a} + \mathbf{b} \rangle = \langle T, T' \rangle = 0.$$

$\square$

In 1858 Gustav Kirchhoff realized (cf. [19]) that in the form (ii) or (iii) the equations for T describe the motion of the axis of a heavy symmetric top (or gyroscope). The vector $\mathbf{a}$ describes the direction of gravity and μ is related to the spinning speed of the gyroscope (see Fig. 5.4). Figure 5.5 illustrates an interesting special case of the characterization given in part (vi). of Theorem 5.7: Assume $\mathbf{b} = 0$, $\mathbf{a} \neq 0$ and

$$\langle \mathbf{e}_3, \mathbf{a} \rangle = \langle \mathbf{e}_3, \gamma(a) \rangle = \langle \mathbf{e}_3, \gamma'(a) \rangle = 0.$$

Then equation in (vi) can be written as

$$\gamma'' = \langle \gamma', \mathbf{a} \rangle \gamma - \langle \gamma', \gamma \rangle \mathbf{a}$$

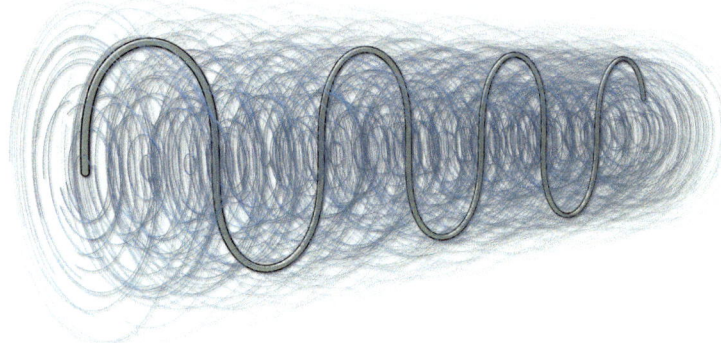

Fig. 5.5 A unit speed elastic curve γ can be described as the orbit of a charged particle moving in a linear magnetic field $p \mapsto B(p) = a \times p + b$. If the initial velocity $T(a)$ is orthogonal to $B(\gamma(a))$, the elastic curve lies in a plane

and therefore the function

$$g \colon [a, b] \to \mathbb{R}, \; g = \langle \mathbf{e}_3, \gamma \rangle$$

satisfies the linear second order equation

$$g'' = \langle \gamma', \mathbf{a} \rangle g$$

with the initial condition $g(a) = g'(a) = 0$. It follows that g vanishes identically and the image of γ is contained in the plane $\mathbb{R}^2 \subset \mathbb{R}^3$ given by $E = \{(x, y, z) \in \mathbb{R}^3 \mid z = 0\}$. Assuming that γ has unit speed, T is the unit tangent of γ and κ its curvature, we can rewrite the equation in (vi) further as

$$\kappa J T = \langle T, \mathbf{a} \rangle \gamma - \langle T, \gamma \rangle \mathbf{a}$$

$$= \frac{\langle \gamma, J\mathbf{a} \rangle}{\langle \mathbf{a}, \mathbf{a} \rangle} (\langle T, \mathbf{a} \rangle J\mathbf{a} - \langle T, J\mathbf{a} \rangle \mathbf{a})$$

$$= \langle \gamma, J\mathbf{a} \rangle J T.$$

The second of the above equalities can be verified by expanding γ as

$$\gamma = \left\langle \gamma, \frac{\mathbf{a}}{|\mathbf{a}|} \right\rangle \frac{\mathbf{a}}{|\mathbf{a}|} + \left\langle \gamma, J\frac{\mathbf{a}}{|\mathbf{a}|} \right\rangle J\frac{\mathbf{a}}{|\mathbf{a}|}$$

while the third equality follows by a similar expansion of T. This means that $\kappa(x)$ is proportional to the distance of $\gamma(x)$ to the line in $\mathbb{R}^2$ through the origin with direction $\frac{\mathbf{a}}{|\mathbf{a}|}$:

$$\kappa = \langle \gamma, J\mathbf{a} \rangle.$$

Amazingly, this is exactly the description of planar elastic curves that had been given by Jakob Bernoulli in 1691 (see [25] for a historical survey, or [14]).

5.3 Vortex Filament Flow

Vortex filaments are curves of singularly concentrated vorticity in a moving fluid. Familiar examples are tornados and smoke rings. The mathematical theory of vortex filament motion started with Lord Kelvin, who in 1880 investigated the evolution of small pertubations of a straight vortex filament (cf. [39]). Later, these perturbations were called Kelvin waves. The full evolution equation for thin vortex filaments was found in 1906 by Tullio Levi-Civita and his student Luigi Sante da Rios. For a detailed history see [34]. Mathematically, the motion of a vortex filament can be described by a one-parameter family γ_t of curves and the da Rios equation says that this one-parameter family satisfies (Fig. 5.6)

$$\dot{\gamma}_t = \frac{d\gamma_t}{ds} \times \frac{d^2\gamma_t}{ds^2}.$$

Another breakthrough occurred in 1972 (cf. [15]) when Hidenori Hasimoto showed that the da Rios equation is equivalent to the non-linear Schrödinger equation:

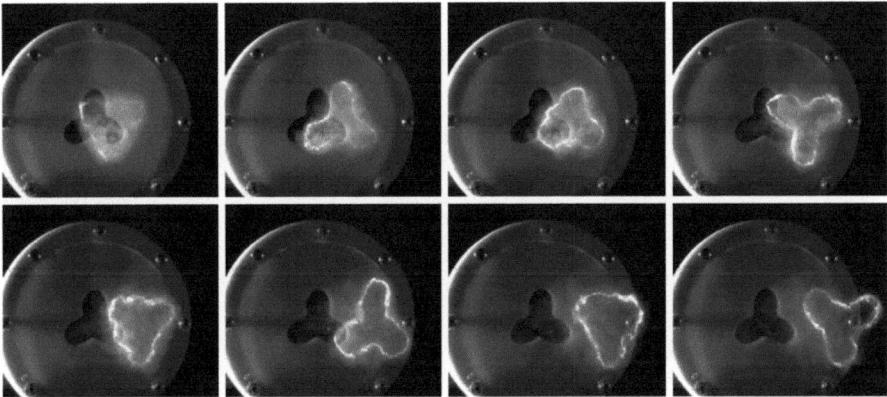

Fig. 5.6 A curve evolving according to the da Rios equation adapted from [20] with permission from William Irvine

> **Theorem 5.8**
>
> *Let $t \mapsto \gamma_t$ with $t \in [t_1, t_2]$ be a smooth one-parameter family of unit speed curves $\gamma_t : [a, b] \to \mathbb{R}^3$ which solves the da Rios equation. Let T_t be the unit tangent field of γ_t. Then there is a smooth family $t \mapsto W_t$ of unit vectors $W_t \in T_t(a)^\perp$ such that the corresponding family $t \mapsto \kappa_t$ of curvature functions for the curves γ_t satisfy*
>
> $$J\dot{\kappa}_t + \kappa_t'' + \frac{|\kappa_t|^2}{2}\kappa_t = 0.$$

Proof. We choose an arbitrary unit vector $\hat{W} \in T_{t_1}(a)^\perp$ and define for $t \in [t_1, t_2]$ a family of unit vectors $W_t \in T_t(a)^\perp$ as the solution of the linear initial value problems

$$W_{t_1} = \hat{W}$$
$$\dot{W}_t = -\langle W_t, \dot{T}_t(a)\rangle T_t(a) - \frac{1}{2}\langle T_t'(a), t_t'(a)\rangle T_t(a) \times W_t.$$

Let Z_t be the parallel normal field along γ_t with $Z_t(a) = W_t$. By Theorem 5.4

$$\left(\langle \dot{Z}_t, T_t \times Z_t\rangle + \frac{1}{2}\langle T_t', T_t'\rangle\right)' = \langle T_t, \dot{T}_t \times T_t'\rangle + \langle T_t', T_t''\rangle$$
$$= \langle T_t, \left(T_t \times T_t''\right) \times T_t'\rangle + \langle T_t', T_t''\rangle$$
$$= 0,$$

where we used that the assumption that γ_t solves the da Rios equation for all t implies

$$\dot{T} = T \times T''.$$

By construction we have $W_t = Z_t(a)$, hence

$$\left(\langle \dot{Z}_t, T_t \times Z_t\rangle + \frac{1}{2}\langle T_t', T_t'\rangle\right)(a) = 0$$

and therefore

$$\langle \dot{Z}_t, T_t \times Z_t\rangle = -\frac{1}{2}\langle T_t', T_t'\rangle.$$

Using the formulas for T'' and T''' from Sect. 4.3 and Theorem 5.5, it then follows that

$$N\left(\dot{\kappa}_t - \frac{|\kappa_t|^2}{2} J\kappa_t\right) \equiv (N\kappa_t)^\bullet$$

$$= -(T')^\bullet$$

$$= -(\gamma_t'')^\bullet$$

$$= -\dot{\gamma}_t''$$

$$= -T' \times T'' - T \times T'''$$

$$\equiv |\kappa_t|^2 T' \times T - T \times N(|\kappa_t|^2 \kappa_t - \kappa_t'')$$

$$= T \times (N\kappa_t'')$$

$$= N(J\kappa_t'') \mod T,$$

where N is again the matrix of parallel normal fields which is used for the definition of κ. This implies that $t \mapsto \kappa_t$ satisfies the nonlinear Schrödinger equation. □

The nonlinear Schrödinger equation was known to be a so-called Soliton equation, and as a consequence also the da Rios equation admits infinitely many constants of the motion. Finally, in 1983 Marsden and Weinstein established (cf. [28]) vortex filament motion as a Hamiltonian mechanical system in its own right (see [10] for a survey article).

The closed curve in Fig. 3.4 is a critical point of bending energy under the constraint of fixed length and fixed enclosed area (cf. [2, Figure 8]) . It can be shown that this curve is the initial curve γ_0 of a solution $t \mapsto \gamma_t$ of the da Rios equation that is defined for all times $t \in \mathbb{R}$. In fact, it can also be proven that this solution is periodic in t. This solution (shown in Fig. 5.7) matches quite well the qualitative behavior of the vortex filament shown in Fig. 5.6.

Definition 5.9

A vector field $X \colon \mathbb{R}^3 \to \mathbb{R}^3$ is called an **infinitesimal rigid motion** if there are vectors $\mathbf{a}, \mathbf{b} \in \mathbb{R}^3$ such that for all $\mathbf{p} \in \mathbb{R}^3$ we have

$$X(\mathbf{p}) = \mathbf{a} \times \mathbf{p} + \mathbf{b}.$$

The following is a reformulation of part (v) of Theorem 5.7:

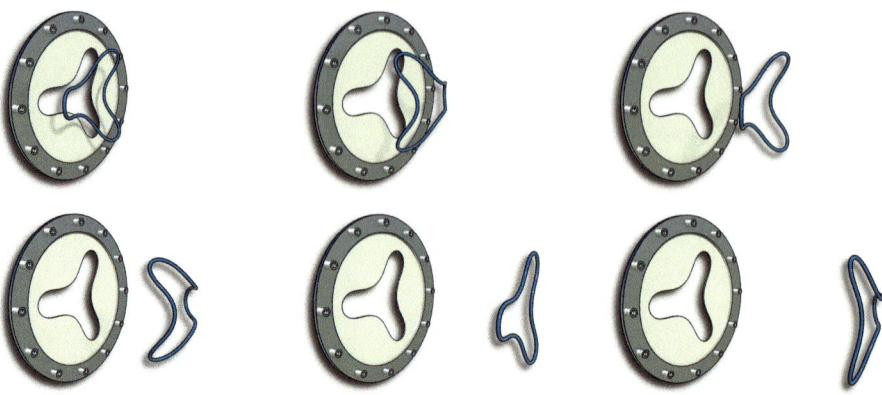

Fig. 5.7 An initial curve which evolves according to the da Rios equation matches the qualitative behavior of a vortex filament (cf. Fig. 5.6). This is why the resulting flow is also referred to as **"vortex filament flow"**, or **"smoke ring flow"**

Fig. 5.8 By Theorem 5.7, an elastic curve that evolves according to the da Rios equation will just undergo rigid motions. These rigid motions can be pure translations *(left)*, pure rotations *(right)* or screw motions *(middle)*

Theorem 5.10

A curve $\gamma : [a, b] \to \mathbb{R}^3$ is elastic if and only if there is an infinitesimal rigid motion X such that for every point of the curve the velocity $\dot{\gamma}$ prescribed by the da Rios equation is given by evaluating X at that point:

$$\dot{\gamma} = X \circ \gamma.$$

Figure 5.8 shows closed elastic curves and their evolution under the da Rios equation.

5.4   Total Squared Torsion

For pioneers of Differential Geometry like Jakob Bernoulli and Leonard Euler (cf. [25] for a historical survey), the motivation for studying elastic curves was to determine the shape γ of a perfectly elastic thin wire (originally shaped as a straight line segment of fixed length when it came out of the factory). In a stable equilibrium position of such a wire the elastic energy stored in its deformation is minimized. Since Bernoulli (1691) and Euler (1744) focused on plane curves, bending energy as introduced in Sect. 1.3 was sufficient for modeling elastic energy.

Later, Lagrange (1788, cf. [22]) and Binet (1844, cf. [5]) realized that for wires in space bending energy alone does not account for all relevant contributions to elastic energy. See also [40] and [4].

In $\mathbb{R}^3$ one has to take account of the internal twisting of the wire: imagine a parallel normal field marked as a colored line on the surface on the wire in its original straight shape. If we bend the wire into space and want to know the elastic energy stored in the deformation, it is not enough to know the resulting shape γ of the wire. We also have to know where the colored line goes on the deformed curve. The twisting of the wire made visible by the colored line contributes to the elastic energy, even if the curve is not changed at all (cf. Fig. 5.9). This means that elastic wires in $\mathbb{R}^3$ are more adequately modelled as a framed curve:

Definition 5.11

A **framed curve** in $\mathbb{R}^3$ is a curve $\gamma : [a, b] \rightarrow \mathbb{R}^3$ together with a unit normal field N along γ.

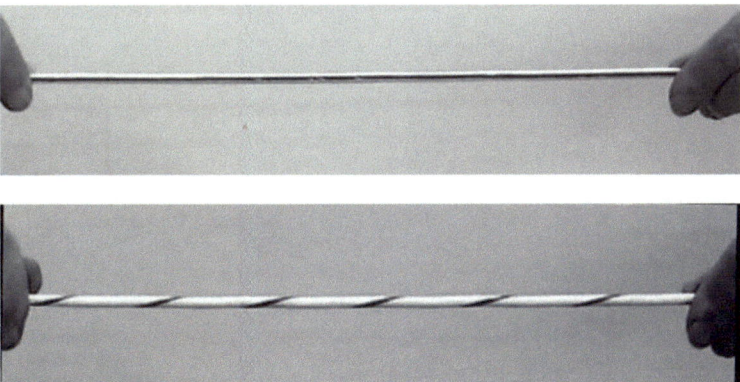

Fig. 5.9 *Top:* An elastic wire with no twist, as indicated by the red line. *Bottom:* The same wire in the same shape, only twisted (Reproduced from [14] with permission from Geoff Goss)

Instead of drawing many arrows, we will usually indicate the unit normal field N of a framed curve by marking a colored line on a slightly thickened version of the curve.

Definition 5.12

For a unit normal field N along a curve $\gamma : [a, b] \rightarrow \mathbb{R}^3$, the function $\tau : [a, b] \rightarrow \mathbb{R}$ given by

$$\tau = \left\langle \frac{dN}{ds}, T \times N \right\rangle$$

is called the **torsion** of N.

The torsion τ measures the deviation of N from being a parallel normal field. After choosing unit vectors $W_a \in T(a)^\perp$ and $W_b \in T(b)^\perp$, we can assign a **total torsion angle** also to a framed curve:

$$\mathcal{T}_W(\gamma, N) := \beta - \alpha$$

where the angles $\alpha, \beta \in \mathbb{R}/2\pi\mathbb{Z}$ are defined by

$$N(a) = \cos\alpha \ W_a + \sin\alpha \ T(a) \times W_a$$
$$N(b) = \cos\beta \ W_b + \sin\beta \ T(b) \times W_b.$$

$\mathcal{T}_W(\gamma, N)$ is related to the total torsion $\mathcal{T}_W(\gamma)$ of the curve γ itself as follows:

Theorem 5.13
Let N be a unit normal field with torsion τ along a curve $\gamma : [a, b] \rightarrow \mathbb{R}^3$ with unit tangent field T. Then, for any choice of unit vectors $W_a \in T(a)^\perp$ and $W_b \in T(b)^\perp$ we have

$$\mathcal{T}_W(\gamma, N) \equiv \mathcal{T}_W(\gamma) + \int_a^b \tau \, ds \qquad \mod 2\pi\mathbb{Z}.$$

Proof. Let Z be the parallel normal field along γ with $Z(a) = W_a$. Then there is a unique function $\eta : [a, b] \rightarrow \mathbb{R}$ with $\eta(a) = \alpha$ such that

$$N = \cos\eta \ Z + \sin\eta \ T \times Z.$$

We have

$$\tau = \left\langle \frac{dN}{ds}, T \times N \right\rangle$$

$$= \left\langle -\frac{d\eta}{ds} \sin\eta \ Z + \frac{d\eta}{ds} \cos\eta \ T \times Z, \ -\sin\eta \ Z + \cos\eta \ T \times Z \right\rangle$$

$$= \frac{d\eta}{ds}.$$

Furthermore,

$$Z(b) = \cos \mathcal{T}_W(\gamma) \ W_b + \sin \mathcal{T}_W(\gamma) \ T(b) \times W_b$$

and therefore

$$\cos\beta \ W_b + \sin\beta \ T(b) \times W_b$$
$$= N(b)$$
$$= \cos\eta(b) \ Z(b) + \sin\eta(b) \ T(b) \times Z(b)$$
$$= \cos\left(\mathcal{T}_W(\gamma) + \eta(b)\right) \ W_b + \sin\left(\mathcal{T}_W(\gamma) + \eta(b)\right) \ T(b) \times W_b.$$

This means that

$$\mathcal{T}_W(\gamma, N) + \alpha \equiv \beta$$
$$\equiv \mathcal{T}_W(\gamma) + \eta(b)$$
$$\equiv \mathcal{T}_W(\gamma) + \eta(a) + \int_a^b \frac{d\eta}{ds}$$
$$\equiv \mathcal{T}_W(\gamma) + \alpha + \int_a^b \tau.$$

$\square$

Theorem 5.13 also explains why we use the terminology "total torsion".

Figure 5.9 (taken from [14]) shows on the bottom a configuration where the end points of the wire are still the same as in the relaxed configuration and therefore, in view of the fixed length, the curve γ is still a straight line segment. However, additional energy has been stored in the twisting of the frame. In Fig. 5.10, moving the end points closer together has made it possible to for the wire to move away from the shape that would minimize bending energy in order to reduce its internal twisting.

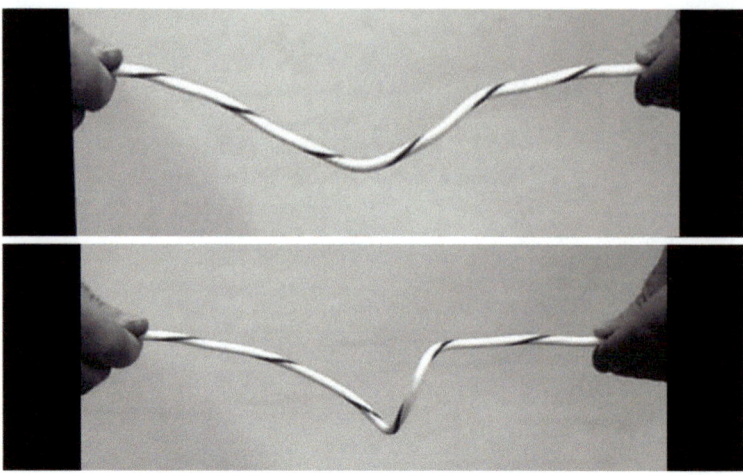

Fig. 5.10 The same wire as in Fig. 5.9. The wire is still twisted by the same amount, but the endpoints are moved closer together (Reproduced from [14] with permission from Geoff Goss)

One can show that in the limit of thin wires (where the thickness tends to zero) this additional energy is of the form

$$c\,\mathcal{S}(\gamma, N),$$

where c is a positive constant and $\mathcal{S}(\gamma, N)$ is defined as follows:

Definition 5.14

Let N be a unit normal field with torsion τ along a curve $\gamma : [a, b] \to \mathbb{R}^3$. Then the **total squared torsion** of the framed curve (γ, N) is defined as

$$\mathcal{S}(\gamma, N) = \frac{1}{2} \int_a^b \tau^2 \, ds.$$

The constant c depends on material properties and on the thickness r of the wire. Following [37], we call c the **twisting modulus**. We work in units where the bending energy is given as in Definition 1.19. Starting from the formulas for the restoring torque and the bending stiffness, one finds that

$$c = \frac{G}{E} = \frac{1}{2(1 + \nu)}$$

where G is the **shear modulus** of the wire material, E is the **Young modulus** and ν is the **Poisson ratio**. According to a table (cf. [9]) of Poisson ratios for common materials, the dimensionless constant c lies between $\frac{1}{3}$ and $\frac{1}{2}$. For example, copper

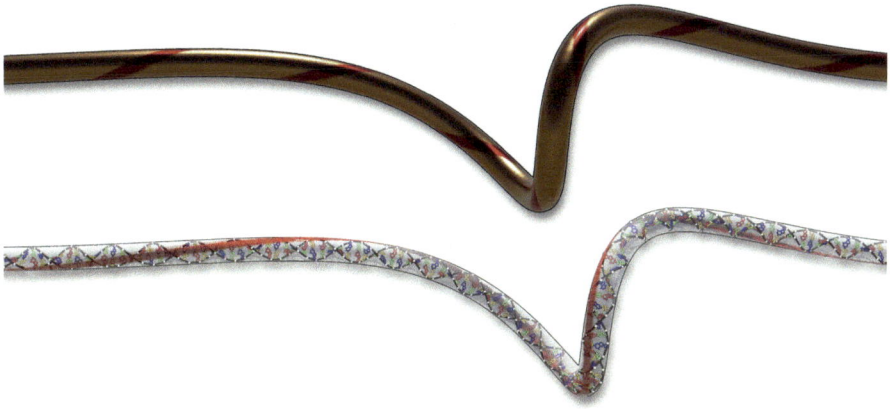

Fig. 5.11 Due to the different twisting moduli, different amounts of torsion are needed to form the same curve shape from a copper wire (*top*, $c = \frac{3}{8}$), or a DNA-strand (*bottom*, $c = \frac{9}{5}$)

wires have $c = \frac{3}{8}$. Also the twisting and bending of DNA strands (where there is no real "material") can be modeled in the same way, see equation 4.1 of [37]. Depending on the ambient conditions, we have $\frac{1}{2} \le c \le 2$ (see Fig. 5.11).

Fortunately, as we will see in Sect. 5.5, the specific value of c is irrelevant for the possible shapes of elastic curves, i.e. of those curves γ of a given length that are critical points of the total elastic energy $\mathcal{B} + \mathcal{S}$. The value of c only effects the normal field N that goes together with such a curve γ, not the shape of γ itself.

5.5   Elastic Framed Curves

In Sect. 5.4 we looked at the elastic energy (including the part that is due to internal twisting) stored in a perfectly elastic wire (modeled as a framed curve (γ, N)) that came out of the factory as a straight line segment. Here we will show that the for an energetic equilibrium configuration of such a wire the curve γ is an elastic curve (Definition 5.6) and the torsion τ of the unit normal field N is constant.

Definition 5.15

Let $\gamma : [a, b] \to \mathbb{R}^3$ be a curve and N a unit normal field along γ. Then a smooth one-parameter family $t \mapsto (\gamma_t, N_t)$ of framed curves is called a **variation with support in the interior** of $[a, b]$ if $\gamma_t(x) = \gamma(x)$ and $N_t(x) = N(x)$ for all x near the end points of the interval $[a, b]$.

Definition 5.16

A framed curve (γ, N) is called an **elastic framed curve with twisting modulus**
$c > 0$ if

$$\frac{d}{dt}\Big|_{t=0} (\mathcal{B}(\gamma_t) + c\, \mathcal{S}(\gamma_t, N_t)) = 0$$

for all variations $t \mapsto (\gamma_t, N_t)$ of (γ, N) with support in the interior of $[a, b]$
which fix the length, i.e. for which

$$\frac{d}{dt}\Big|_{t=0} \mathcal{L}(\gamma_t) = 0.$$

Which curves γ in $\mathbb{R}^3$ can be supplemented by a unit normal field N in such a
way that (γ, N) is an elastic framed curve? It turns out that those curves are precisely
the elastic curves:

Theorem 5.17
*A framed curve (γ, N) in $\mathbb{R}^3$ is elastic with twisting modulus c if and only if
its torsion τ is constant and γ is a critical point of*

$$\mathcal{B} + c\tau\, \mathcal{T}$$

under the constraint of fixed length $\mathcal{L}$.

Proof. Let (γ, N) be an elastic framed curve elastic with twisting modulus c. Let us
first consider special variations $t \mapsto (\gamma_t, N_t)$ of (γ, N) with support in the interior
of $[a, b]$ for which the curve itself does not move at all, i.e. for all t we have $\gamma_t = \gamma$,
so that for those variations we have

$$\frac{d}{dt}\Big|_{t=0} \mathcal{B}(\gamma_t) = 0.$$

The normals that we consider are of the form

$$N_t = \cos(t\alpha)N + \sin(t\alpha)T \times N.$$

where $\alpha\colon [a, b] \to \mathbb{R}$ is a function with support in the interior of $[a, b]$. Then
$\dot{N} = \alpha T \times N$, $\dot{T} = 0$, $\dot{ds} = 0$ and

$$\dot{\tau} = \left\langle \frac{dN}{ds}, T \times N \right\rangle^{\bullet}$$

$$= \left\langle \frac{d\dot{N}}{ds}, T \times N \right\rangle + \left\langle \frac{dN}{ds}, T \times \dot{N} \right\rangle$$

$$= \frac{d\alpha}{ds} \langle T \times N, T \times N \rangle + \alpha \left\langle \frac{dN}{ds}, T \times (T \times N) \right\rangle$$

$$= \frac{d\alpha}{ds}.$$

Therefore, for all such functions α we have

$$0 = \frac{d}{dt}\bigg|_{t=0} (\mathcal{B}(\gamma_t) + c\, \mathcal{S}(\gamma_t, N_t))$$

$$= c \frac{d}{dt}\bigg|_{t=0} \int_a^b \frac{\tau^2}{2} ds$$

$$= c \int_a^b \tau \dot{\tau}\, ds$$

$$= c \int_a^b \tau \frac{d\alpha}{ds} ds$$

$$= c \int_a^b \tau \alpha'$$

$$= -c \int_a^b \tau' \alpha$$

and therefore we must have $\tau' = 0$. Let now $t \mapsto \gamma_t$ be an arbitrary variation of γ with support in the interior of $[a, b]$ for which

$$\frac{d}{dt}\bigg|_{t=0} \mathcal{L}(\gamma_t) = 0.$$

Then, for small t, we can define unit normal fields N_t along γ_t (equal to N near the end points of the interval $[a, b]$) by projecting $N(x)$ to $T_t(x)^\perp$ where T_t is the unit tangent field of γ_t:

$$N_t := \frac{N - \langle N, T_t \rangle T_t}{|N - \langle N, T_t \rangle T_t|}.$$

Then, by Theorem 5.13 and with $\frac{d}{dt}\big|_{t=0}\mathcal{L}(\gamma_t) = 0$, we have

$$
\frac{d}{dt}\bigg|_{t=0} (\mathcal{B} + c\tau\,\mathcal{T})\,(\gamma_t, N_t) = \frac{d}{dt}\bigg|_{t=0} \mathcal{B}(\gamma_t) + c\tau \int_a^b \dot{\tau}\,ds
$$

$$
= \frac{d}{dt}\bigg|_{t=0} \mathcal{B}(\gamma_t) + c\int_a^b \tau\dot{\tau}\,ds + c\frac{\tau^2}{2}\dot{\mathcal{L}}
$$

$$
= \frac{d}{dt}\bigg|_{t=0} (\mathcal{B}(\gamma_t) + c\,\mathcal{S}(\gamma_t, N_t))
$$

$$
= 0.
$$

This proves the "only if" direction of our claim. We leave the "if" direction to the reader. $\square$

5.6   Frenet Normals

Definition 5.18

A unit normal field $N : [a, b] \to \mathbb{R}^n$ along a curve $\gamma : [a, b] \to \mathbb{R}^n$ with unit tangent T is called a **Frenet normal** field if there is a function $\kappa_f : [a, b] \to \mathbb{R}$ such that

$$
\frac{dT}{ds} = -\kappa_f N.
$$

If we ignore the effects of gravity, the unit vector pointing upward in the reference frame of an airplane like the one in Fig. 5.12 (which is lacking a rudder) will be a Frenet normal for its flight path (see Fig. 5.13). A curve $\gamma : [a, b] \to \mathbb{R}^2$ has exactly two unit normal fields, and both of them are Frenet. The one with $N = -JT$ has $\kappa_f = \kappa$ where κ is the curvature function of γ.

Fig. 5.12 The vertical vector in the reference frame of an airplane with no rudder is a Frenet normal along its flight path

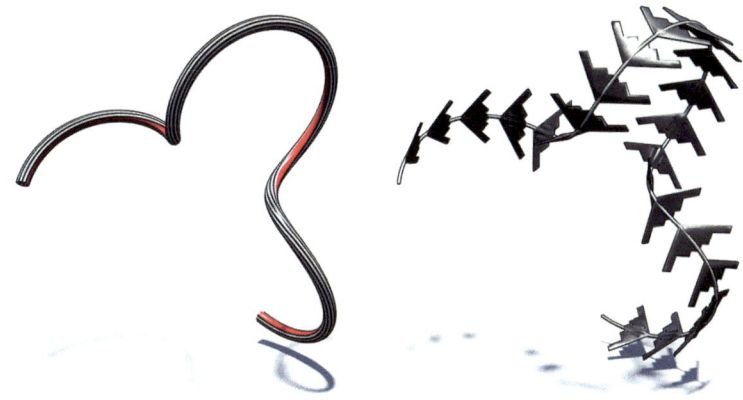

Fig. 5.13 A Frenet normal along a space curve

Not every curve in $\mathbb{R}^3$ has a Frenet normal field. For example, any Frenet normal field N for the curve (cf. [36, Chapter 1])

$$\gamma : [-1, 1] \to \mathbb{R}^3, \ \gamma(x) = \begin{cases} (x, e^{\frac{1}{x}}, 0), & x < 0 \\ (0, 0, 0), & x = 0 \\ (x, 0, e^{-\frac{1}{x}}), & x > 0 \end{cases}$$

would have to satisfy (see Fig. 5.14)

$$N(x) = \begin{cases} \pm \mathbf{e}_3 & x < 0 \\ \pm \mathbf{e}_2 & x > 0, \end{cases}$$

which is impossible for a smooth map. Figure 5.14 shows a curve where four planar curves are stitched together in a smooth fashion, together with an attempt to define a Frenet normal field for this curve.

Even if a Frenet normal field exists on an open dense set of $[a, b]$ (which in general is not guaranteed), it can exhibit singularities that can be worse than the jump discontinuities from the previous example. For example, any Frenet normal field for the curve

$$\gamma : [-1, 1] \to \mathbb{R}^3, \ \gamma(x) = \begin{cases} (x, e^{\frac{1}{x}} \cos(\frac{1}{x}), e^{\frac{1}{x}} \sin(\frac{1}{x})), & x < 0 \\ (t, 0, 0), & x \geq 0. \end{cases}$$

will have unbounded rotation speed, as is visible in Fig. 5.15. After Frenet normals were introduced in the middle of the nineteenth century, they quickly became a popular tool for studying curves. The second half of the nineteenth century saw

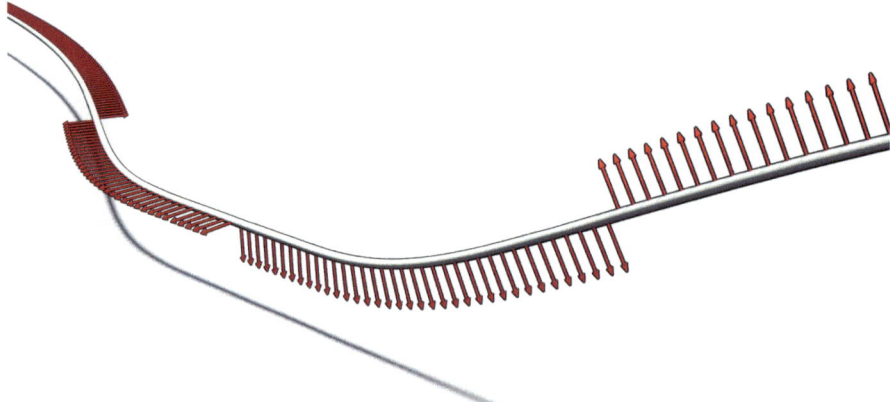

Fig. 5.14 Even when a Frenet normal field exists on an open dense set, on the whole curve there might be no such field

Fig. 5.15 Away from a single point, this curve has a Frenet normal. However, its rotation speed τ is unbounded

the powerful appearance of Complex Analysis and Algebraic Geometry in the landscape of Mathematics, while Topology (and certainly Differential Topology) were still in their infancy. In those days it seemed natural to assume that the curves γ under consideration were real analytic (locally representable as a power series). And every real analytic curve does indeed have a Frenet normal field:

Theorem 5.19
If $\gamma : [a, b] \to \mathbb{R}^n$ is real analytic, then γ has a Frenet frame.

Proof. Without loss of generality we may assume that γ has unit speed. If γ parametrizes a piece of a straight line, then every unit normal field along γ is Frenet and we are done. Otherwise, because of the real analyticity of γ, there are only finitely many parameter values $x_1, \ldots x_m \in [a, b]$ where γ'' vanishes. On each subinterval of $[a, b]$ bounded by two of the points $a, x_1, \ldots, x_m, b$ there is a Frenet normal field, unique up to sign, which is obtained by setting $N = \frac{\gamma''}{|\gamma''|}$. It is therefore sufficient to show that also in the neigborhood of each x_j there is a Frenet normal field, unique up to sign. In the end, the signs can then easily be adjusted to yield a Frenet normal field on the whole interval $[a, b]$. By real analyticity, there is a neighborhood of x_j where γ can be expressed as

$$\gamma(x) = \sum_{k=0}^{\infty} a_k (x - x_j)^k$$

with $a_k \in \mathbb{R}^n$. Then

$$\gamma''(x) = \sum_{k=2}^{\infty} k(k-1) a_k (x - x_j)^{k-2}$$

and there is an index $\ell \in \mathbb{N}$ such that $a_k = 0$ for $k = 2, \ldots, \ell - 1$ but $a_\ell \neq 0$. Then

$$\gamma'' = (x - x_j)^{\ell-2} \sum_{k=0}^{\infty} (\ell+k)(\ell+k-1) a_{\ell+k} (x - x_j)^k =: (x - x_j)^{\ell-2} \eta(x)$$

with $\eta(x) \neq 0$ for all x in some neighborhood of x_j. In this neighborhood

$$N(s) := \frac{\eta}{|\eta|}.$$

is the desired Frenet normal field. □

Nowadays, the standard assumption for curves is that they are smooth, i.e. infinitely often differentiable. Because for $n \geq 3$ not every C^∞ curve in $\mathbb{R}^n$ has a Frenet normal field, for $n \geq 3$ these fields cannot be used for studying global questions about smooth curves in $\mathbb{R}^n$. Moreover, when used in the context of numerical algorithms that operate on space curves, Frenet normals can cause unexpected behavior near curves that do not have a Frenet normal.

Part II

Surfaces

Surfaces and Riemannian Geometry

6

The most simple quantity of a one-dimensional curve $\gamma: [a, b] \to \mathbb{R}^n$ is its speed $|\gamma'|: [a, b] \to \mathbb{R}$. The goal of this chapter is to arrive at the analogous statement for a two-dimensional surface in $\mathbb{R}^n$. Our first task will be to replace the interval $[a, b]$ by a suitable domain of definition $M \subset \mathbb{R}^2$. For a surface $f: M \to \mathbb{R}^n$ the analog for the speed of a curve will be a *Riemannian metric* induced on M by f.

6.1   Surfaces in $\mathbb{R}^n$

Our investigations of curves in $\mathbb{R}^n$ will be the guideline when we now start to study surfaces. For the most part we will focus on surfaces in $\mathbb{R}^3$. We will study the curvature of surfaces and the analog of the length of a curve (obviously the area of a surface) as well as the analog of the total squared curvature (called the Willmore functional). We will study the critical points of the area under variations with support in the interior (these surfaces are called minimal surfaces) and of the Willmore functional. We will prove a famous result that concerns the surface analog of $\int_a^b \kappa \, ds$, the so-called Gauss-Bonnet theorem. We will investigate the analog (called the Euler characteristic) for the tangent winding number of a curve in $\mathbb{R}^2$.

In our discussion of (non-closed) curves γ in $\mathbb{R}^n$, γ was always defined on a closed interval $[a, b]$. It would have made little difference if γ would have been defined on the finite union of pairwise disjoint intervals. In the case of surfaces, it will be useful to allow for such disconnected domains.

© The Author(s) 2024
U. Pinkall, O. Gross, *Differential Geometry*, Compact Textbooks in Mathematics,
https://doi.org/10.1007/978-3-031-39838-4_6

Definition 6.1

A subset $M \subset \mathbb{R}^2$ is called a **connected compact domain with smooth boundary** if

$$M = M_0 \setminus \{\mathring{M}_1 \cup \ldots \cup \mathring{M}_k\}$$

where for each $j \in \{0, \ldots, k\}$

$$M_j = \varphi_j(D)$$

is the image of the unit disk

$$D := \{p \in \mathbb{R}^2 \mid |p| \leq 1\}$$

under a diffeomorphism

$$\varphi_j \colon D \to \mathbb{R}^2$$

and the M_j are pairwise disjoint and contained in the interior of M_0. A finite disjoint union of connected compact domains with smooth boundary is called a **compact domain with smooth boundary** (see Fig. 6.1).

Notation Throughout the rest of the book, M will denote a compact domain with smooth boundary in $\mathbb{R}^2$.

In order to avoid having to mention regularity constantly, we include regularity in the definition of a surface (see Fig. 6.2):

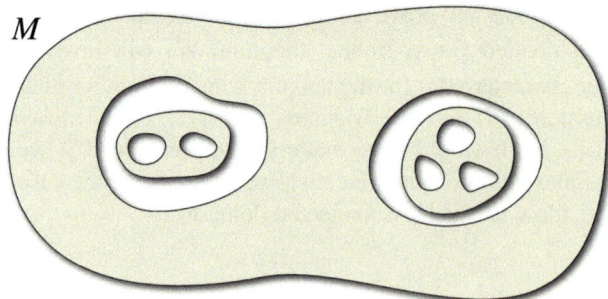

Fig. 6.1 A compact domain with smooth boundary

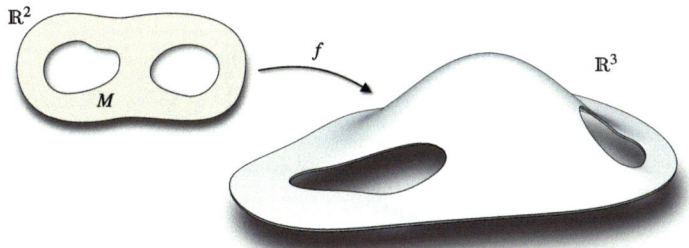

Fig. 6.2 A surface $f: M \to \mathbb{R}^3$

Definition 6.2

A **surface** in $\mathbb{R}^n$ is a smooth map $f: M \to \mathbb{R}^n$ whose derivative $f'(p)$ is an $(n \times 2)$-matrix of rank 2 for all $p \in M$.

We will denote the coordinates in $\mathbb{R}^2$ by u and v. Partial derivatives with respect to u or v will be denoted by subscripts, so for a surface f in $\mathbb{R}^n$ the matrix-valued function $f': M \to \mathbb{R}^{n \times 2}$ is of the form

$$f' = \begin{pmatrix} | & | \\ f_u & f_v \\ | & | \end{pmatrix}$$

with $f_u(p), f_v(p) \in \mathbb{R}^n$ linearly independent for all $p \in M$. The following two definitions are special cases of the ones in Appendix A.1.

Definition 6.3

A map $f: M \to \mathbb{R}^n$ is called smooth if it is of the form $f = \tilde{f}|_M$ for some smooth map $\tilde{f}: U \to \mathbb{R}^n$ where $U \subset \mathbb{R}^2$ is an open set that contains M.

It is easy to check that even at boundary points $p \in M$ the Jacobian matrix $\tilde{f}'(p)$ only depends on f, not on the specific way in which $\tilde{f}$ extends f. We therefore can safely define $f'(p) := \tilde{f}'(p)$.

Definition 6.4

If $M, \tilde{M} \subset \mathbb{R}^2$ are two compact domains with smooth boundary, a bijective map $\varphi: M \to \tilde{M}$ is called a **diffeomorphism** if both φ and φ^{-1} are smooth. A diffeomorphism φ is called **orientation-preserving** if $\det \varphi'(p) > 0$ for all $p \in M$.

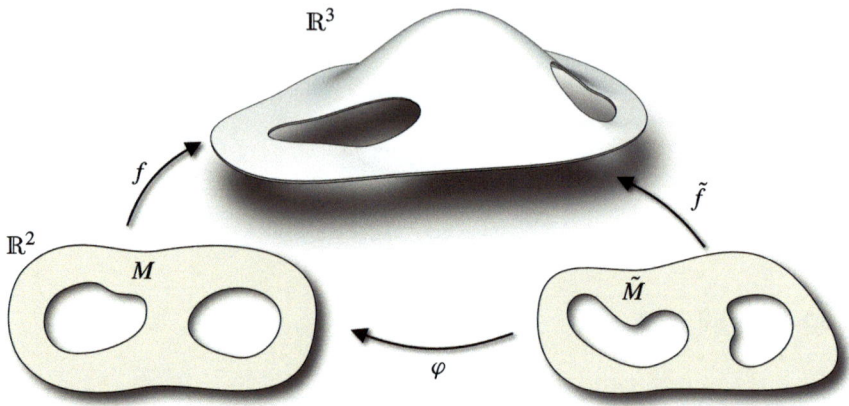

Fig. 6.3   A reparametrization $\tilde{f}$ of a surface f

Definition 6.5

If $f: M \rightarrow \mathbb{R}^n$ and $\tilde{f}: \tilde{M} \rightarrow \mathbb{R}^n$ are two surfaces, then $\tilde{f}$ is called an (orientation-preserving) **reparametrization** of f if there is an (orientation-preserving) diffeomorphism $\varphi: \tilde{M} \rightarrow M$ such that

$$\tilde{f} = f \circ \varphi,$$

(see Fig. 6.3).

As in the case of curves in the plane, it is not difficult to check that reparametrization (as well as orientation-preserving reparametrization) defines an equivalence relation on the set of surfaces in $\mathbb{R}^n$. Although we will not formalize this, we are only interested in properties of surfaces that are invariant under orientation-preserving reparametrization, so the real objects of our study are the equivalence classes of surfaces under reparametrization.

6.2   Tangent Spaces and Derivatives

Let $M \subset \mathbb{R}^2$ be a compact domain with smooth boundary and $f: M \rightarrow \mathbb{R}^k$ a smooth map. Then the directional derivative of f at a point $p \in M$ in the direction of a vector $\hat{X} \in \mathbb{R}^2$ is given by

$$df(p, \hat{X}) := f'(p)\hat{X}.$$

This means that all these directional derivatives are encoded in a map $df : M \times \mathbb{R}^2 \to \mathbb{R}^k$:

Definition 6.6

For a point $p \in M$, a **tangent vector** to M at p is a pair $X = (p, \hat{X})$ where $\hat{X} \in \mathbb{R}^2$. The set

$$T_p M = \{p\} \times \mathbb{R}^2$$

of all these tangent vectors is called the **tangent space** to M at p. We make each $T_p M$ into a two-dimensional real vector space by defining for $X = (p, \hat{X})$, $Y = (p, \hat{Y})$ and $\lambda \in \mathbb{R}$

$$X + Y = (p, \hat{X} + \hat{Y})$$

$$\lambda X = (p, \lambda \hat{X}).$$

The union $TM = M \times \mathbb{R}^2$ of all these tangent spaces is called the **tangent bundle** of M. The map

$$\pi : TM \to M, \ (p, \hat{X}) \mapsto p$$

is called the **projection map** of the tangent bundle.

One immediate benefit of this definition is a more concise notation for derivatives:

Definition 6.7

For a smooth map $f : M \to \mathbb{R}^n$ we define the **derivative** $df : TM \to \mathbb{R}^n$ of f by setting for $X \in TM$, $X = (p, \hat{X})$

$$df(X) = f'(p)\hat{X},$$

(see Fig. 6.4).

The restriction of df to each tangent space $T_p M$ is a linear map from $T_p M$ to $\mathbb{R}^n$.

Definition 6.8

A smooth map $X : M \to TM$ is called a **vector field** if $\pi \circ X = \mathrm{id}_M$, which means that $X(p) \in T_p M$ for all $p \in M$.

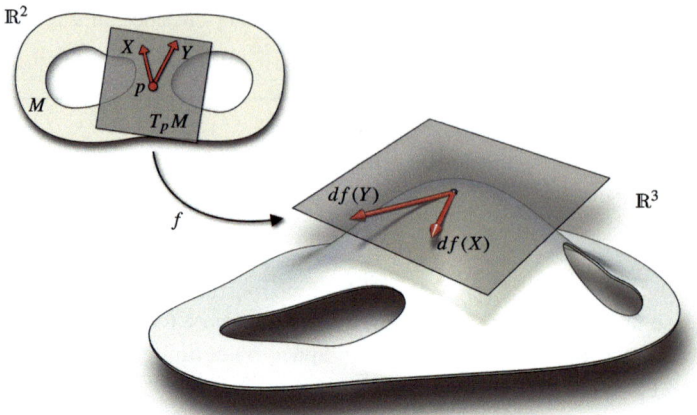

Fig. 6.4 Two tangent vectors $X, Y \in T_p M$ and their image under df

If $\hat{X} \colon M \to \mathbb{R}^2$ is a smooth map, then the assignment

$$X \colon M \to TM, \; X(p) = (p, \hat{X}(p))$$

is a smooth vector field on M and all smooth vector fields are obtained in this way. Here is some convenient notation:

Definition 6.9

 (i) The vector space of all smooth functions $f \colon M \to \mathbb{R}$ is denoted by $C^\infty(M)$.
 (ii) The vector space of all smooth functions $f \colon M \to \mathbb{R}^n$ is denoted by $C^\infty(M, \mathbb{R}^n)$.
 (iii) The vector space of all smooth vector fields on M is denoted by $\Gamma(TM)$.

As is known from calculus class, for a smooth map $f \colon M \to \mathbb{R}^n$ the vector $f'(p)\hat{X} \in \mathbb{R}^n$ can also be interpreted as the directional derivative of f at p in the direction of the vector $\hat{X} \in \mathbb{R}^2$. With this in mind we define the **directional derivative** of $f \in C^\infty(M, \mathbb{R}^n)$ in the direction of a vector field $X \in \Gamma(TM)$ by

$$(d_X f)(p) := d_{X(p)} f = df(X(p)).$$

Definition 6.10

The **coordinate vector fields** $U, V \in \Gamma(TM)$ are defined as

$$U(p) = \left(p, \begin{pmatrix} 1 \\ 0 \end{pmatrix} \right), \qquad V(p) = \left(p, \begin{pmatrix} 0 \\ 1 \end{pmatrix} \right).$$

The directional derivatives in the direction of U or V are just partial derivatives:

$$d_U f = f_u, \qquad d_V f = f_v.$$

Definition 6.11

Let $M, \tilde{M} \subset \mathbb{R}^2$ be two compact domains with smooth boundary and $\varphi: \tilde{M} \to M$ a diffeomorphism. Then we define

$$d\varphi: T\tilde{M} \to TM, \; d\varphi(X) = \left(\varphi(p), \varphi'(p)\hat{X} \right) \quad \text{for} \quad X = \left(p, \hat{X} \right).$$

▶ **Remark 6.12** Note that for $X \in T_p\tilde{M}$ the vector $d\varphi(X)$ is an element of $T_{\varphi(p)}M$, while for a surface $f: M \to \mathbb{R}^n$ the vector $df(X)$ is just an element of $\mathbb{R}^n$, not an element of something like $T_{f(p)}\mathbb{R}^n$. We are relying here on the fact that in our situation we can naturally identify all such tangent spaces $T_q\mathbb{R}^n$ with $\mathbb{R}^n$ itself. This mild context-dependency of notation should lead to no confusion. It is very useful and common in Differential Geometry.

With this notation in place, the chain rule now emerges in its most elegant form:

Theorem 6.13

(i) Suppose $\tilde{f}: \tilde{M} \to \mathbb{R}^n$ is a reparametrization of the surface $f: M \to \mathbb{R}^n$, i.e. $\tilde{f} = f \circ \varphi$ for some diffeomorphism $\varphi: \tilde{M} \to M$. Then

$$d\tilde{f} = df \circ d\varphi.$$

(ii) If $M, \tilde{M}, \hat{M} \subset \mathbb{R}^2$ are compact domains with smooth boundary and $\varphi: \tilde{M} \to M$ and $\tilde{\varphi}: \hat{M} \to \tilde{M}$ are diffeomorphisms, then

$$d(\varphi \circ \tilde{\varphi}) = d\varphi \circ d\tilde{\varphi}.$$

Proof. The proof just involves spelling out our definitions and applying the ordinary chain rule. ☐

6.3   Riemannian Domains

When it comes to investigating the geometry of a surface $f: M \to \mathbb{R}^n$, the geometry of M as it sits in $\mathbb{R}^2$ is completely irrelevant. Things like the length of a vector or the angle between vectors should be computed in the target space $\mathbb{R}^n$ of f, not in $\mathbb{R}^2$. Accordingly, we endow each tangent space $T_p M$ with its own private Euclidean scalar product by defining

$$\langle , \rangle_f : \bigcup_{p \in M} (T_p M \times T_p M) \to \mathbb{R}, \quad \langle X, Y \rangle_f = \langle df(X), df(Y) \rangle.$$

It is easy to check that for each $p \in M$ the restriction of $\langle , \rangle_f$ to $T_p M \times T_p M$ is indeed a positive definite scalar product on $T_p M$. With respect to this scalar product, $X \in T_p M$ is a unit vector if and only if $df(X) \in \mathbb{R}^n$ is a unit vector.

Definition 6.14

$\langle , \rangle_f$ as defined above is called the **metric** on M induced by f.

▶ **Remark 6.15** In older texts the induced metric is often called the first fundamental form. We will not use this terminology.

In general, objects like $\langle , \rangle_f$ are interesting even when they are not induced by a map $f: M \to \mathbb{R}^n$. That is, $\langle , \rangle_f$ has the properties of a scalar product between two vectors, which allows us to measure lengths an angles. However, one can freely choose other ways to define such a metric, without explicit reference to a surface f, as long as the scalar product properties are satisfied.

Definition 6.16

Let M be a compact domain with smooth boundary in $\mathbb{R}^2$. Then:

(i) A map

$$\langle , \rangle : \bigcup_{p \in M} (T_p M \times T_p M) \to \mathbb{R}$$

is called a **Riemannian metric** on M if for each $p \in M$ the restriction of $\langle , \rangle$ to $T_p M \times T_p M$ is a positive definite scalar product and for any two smooth vector fields $X, Y \in \Gamma(TM)$ the function

$$\langle X, Y \rangle : M \to \mathbb{R}$$

is smooth.

(ii) M together with a Riemannian metric $\langle\,,\rangle$ on M is called a **Riemannian domain**.

▶ **Remark 6.17** For brevity of the notation we will usually omit the index $(\cdot)_f$ even for Riemannian metrics which are induced by some $f: M \to \mathbb{R}^3$. The inserted vectors should provide enough context to avoid confusion.

A Riemannian metric $\langle\,,\rangle$ gives rise to a function

$$| \cdot |: TM \to \mathbb{R}, \ |X| = \sqrt{\langle X, X\rangle}.$$

The restriction of $| \cdot |$ to each tangent space is indeed a **norm** on $T_p M$. One should note that the coordinate vector fields U and V are not necessarily orthonormal with respect to this induced metric. In fact, this is only true for special surfaces. We will elaborate more on this in Sect. 6.5.

Example 6.18
The norm corresponding to the metric $\langle\,,\rangle_\iota$ induced by the inclusion map

$$\iota: M \to \mathbb{R}^2, \ (u, v) \mapsto \begin{pmatrix} u \\ v \end{pmatrix}$$

satisfies

$$| \cdot |_\iota^2 = du^2 + dv^2.$$

The above equation should be read as a literal equality of two functions on TM.

6.4   Linear Algebra on Riemannian Domains

Even in the absence of a Riemannian metric, each single tangent space $T_p M$ is a playing field for Linear Algebra.

Definition 6.19

A smooth map

$$A: TM \to TM$$

is called an **endomorphism field** if its restriction to each tangent space $T_p M$ is a linear map

$$A_p: T_p M \to T_p M.$$

Remember that $TM = M \times \mathbb{R}^2$, so it is clear what we mean by a smooth map from TM to TM. In particular, it is clear what we mean by a smooth endomorphism field. For every smooth endomorphism field A there are smooth functions $a, b, c, d \colon M \to \mathbb{R}$ such that

$$AU = a \cdot U + c \cdot V$$

$$AV = b \cdot U + d \cdot V.$$

This means that a smooth endomorphism field basically is the same thing as a smooth map

$$p \mapsto \begin{pmatrix} a & b \\ c & d \end{pmatrix} \in \mathbb{R}^{2 \times 2}.$$

We always have the identity map as a canonical endomorphism field:

$$I \colon TM \to TM, \ IX = X \quad \text{for all } X \in TM.$$

If A is an arbitrary smooth endomorphism field on M, for each $p \in M$ we can take the **determinant** or **trace** of the restriction of A to $T_p M$ and obtain smooth functions

$$\det A, \operatorname{tr} A \colon M \to \mathbb{R}.$$

In the presence of a Riemannian metric we can define the adjoint of an endomorphism field:

> **Theorem 6.20**
>
> *Let $\langle , \rangle$ be a Riemannian metric on M and A a smooth endomorphism field on M. Then there is a unique smooth endomorphism field A^* on M such that for all vector fields $X, Y \in \Gamma(TM)$ we have*
>
> $$\langle AX, Y \rangle = \langle X, A^* Y \rangle.$$

Proof. By definition of a Riemannian metric, the map

$$G \colon M \to \mathbb{R}^{2 \times 2}, \ G = \begin{pmatrix} \langle U, U \rangle & \langle U, V \rangle \\ \langle V, U \rangle & \langle V, V \rangle \end{pmatrix}$$

is smooth and the matrix $G(p)$ is invertible for all $p \in M$. Now one can check that, given an endormorphism field A such that

$$AU = aU + cV$$
$$AV = bU + dV$$

the endomorphism field A^* defined by

$$A^*U = \tilde{a} \cdot U + \tilde{c} \cdot V$$
$$A^*V = \tilde{b} \cdot U + \tilde{d} \cdot V$$

with

$$\begin{pmatrix} \tilde{a} & \tilde{c} \\ \tilde{b} & \tilde{d} \end{pmatrix} = G^{-1} \begin{pmatrix} a & b \\ c & d \end{pmatrix}^T G$$

is smooth and satisfies the desired identity. The uniqueness part of the claim is straightforward.   □

The endomorphism field I defined above is self-adjoint, which means $I^* = I$. The only structure on M we want to inherit from $\mathbb{R}^2$ is the notion of orientation:

Definition 6.21

Two vectors

$$X = (p, \hat{X}) , \quad Y = (p, \hat{Y}) \in T_pM$$

are said to form a **positively oriented basis** of T_pM if $\hat{X}, \hat{Y} \in \mathbb{R}^2$ are a positively oriented basis of $\mathbb{R}^2$, i.e. $\det_{\mathbb{R}^2}(\hat{X}, \hat{Y}) > 0$.

Each tangent space T_pM of a Riemannian domain comes with its own determinant form:

Theorem 6.22

Let $\langle , \rangle$ be a Riemannian metric on M. Then there is a unique map

$$\det: \bigcup_{p \in M} (T_pM \times T_pM) \to \mathbb{R}$$

(continued)

Theorem 6.22 (continued)
such that for every $p \in M$ the restriction

$$\det |_{T_p M \times T_p M}$$

is a skew-symmetric bilinear form such that

$$\det(X, Y) = 1$$

*for every positively oriented orthonormal basis of $T_p M$. The map $\det$ is called the **area form** of the Riemannian domain $(M, \langle\,,\rangle)$.*

Proof. The vector fields

$$X := \frac{U}{\sqrt{\langle U, U\rangle}}$$

$$Y := \frac{\langle U, U\rangle V - \langle V, U\rangle U}{\sqrt{\langle U, U\rangle}\sqrt{\langle U, U\rangle\langle V, V\rangle - \langle U, V\rangle^2}}$$

are orthonormal at each point $p \in M$. Therefore, the function $\det$ we are looking for has to satisfy

$$1 = \det(X, Y) = \frac{\det(U, V)}{\sqrt{\langle U, U,\rangle\langle V, V\rangle - \langle U, V\rangle^2}}.$$

The skew-symmetric bilinear forms on $T_p M$ form a 1-dimensional vector space, so there is a unique such form $\det$ for which

$$\det(U, V) = \sqrt{\langle U, U\rangle\langle V, V\rangle - \langle U, V\rangle^2}\,,$$

where the equality has to be read point-wise. This form already satisfies $\det(X, Y) = 1$. On the other hand, every positively oriented orthonormal basis of $T_p M$ is of the form

$$\tilde{X} = \cos\alpha\, X(p) - \sin\alpha\, Y(p)$$

$$\tilde{Y} = \sin\alpha\, X(p) + \cos\alpha\, Y(p)$$

for some $\alpha \in \mathbb{R}$. Therefore, we also have $\det(\tilde{X}, \tilde{Y}) = 1$. $\square$

Theorem 6.23

Let $\langle\,,\rangle$ be a Riemannian metric on M and det *the area form defined in Theorem 6.22. Then there is a unique endomorphism field J on M such that for all $p \in M$ and all $X, Y \in T_pM$ we have*

$$\langle JX, Y \rangle = \det(X, Y).$$

In terms of the coordinate vector fields, J is given by

$$JZ = \frac{\langle U, Z \rangle V - \langle V, Z \rangle U}{\sqrt{\langle U, U \rangle \langle V, V \rangle - \langle U, V \rangle^2}}.$$

Hence J is smooth. For every positively oriented orthonormal basis of T_pM we have

$$JX = Y$$
$$JY = -X.$$

So, J operates in each tangent space T_pM as the $90°$-rotation in the positive sense.

The proof is straightforward and left to the reader. The theorem will be useful in several instances:

Theorem 6.24

For vectors $X, Y, Z \in T_pM$ the following identity holds:

$$\langle X, Z \rangle Y - \langle Y, Z \rangle X = \det(X, Y) JZ.$$

Proof. Both sides of the claimed identity are linear in X and Y and by Theorem 6.23 the formula that we want to prove is true whenever $X, Y \in \{U(p), V(p)\}$. Since $U(p)$ and $V(p)$ form a basis of T_pM, the claimed identity then is true for all $X, Y \in T_pM$. $\square$

6.5   Isometric surfaces

Definition 6.25

Two surfaces $f, \tilde{f} : M \rightarrow \mathbb{R}^n$ are called **isometric** if they induce the same Riemannian metric $\langle \, , \rangle$ on M.

Note that f and $\tilde{f}$ are isometric if and only if

$$\langle f_u, f_u \rangle = \langle \tilde{f}_u, \tilde{f}_u \rangle$$
$$\langle f_u, f_v \rangle = \langle \tilde{f}_u, \tilde{f}_v \rangle$$
$$\langle f_v, f_v \rangle = \langle \tilde{f}_v, \tilde{f}_v \rangle.$$

The physical intuition concerning isometries is as follows: The deformation of the surface f to the surface $\tilde{f}$ involves only bending, without any intrinsic deformation such as stretching within the surface. In the nineteenth century, geometers liked to demonstrate this using leather patches. By methods known for example to shoemakers, a leather patch can be brought into any initial shape. After the initial preparation, the leather can still be bent easily, but it will not allow stretching. In Fig. 6.5, the initial shape f of the patch is a planar ring. This ring can easily be fitted to a cone, assuming a shape $\tilde{f}$.

It is clear that one can slide the ring freely around on the cone, in all directions. Few surfaces have the property that one can take a piece of the surface and slide it without distortion or stretching around on the surface. For example, the leather patch on the surface in Fig. 6.6 is clearly stuck in place. One famous surface on which such a patch can freely slide, already known in the nineteenth century and a popular tool for the leather demonstration, is the **pseudosphere**, that can be parametrized as follows:

Fig. 6.5 The maps f and $\tilde{f}$ are isometric, as the flat leather patch f can be placed onto a cone without tearing or stretching. It can even slide freely on the cone

Fig. 6.6 The leather patch on this dodecahedron is stuck in place

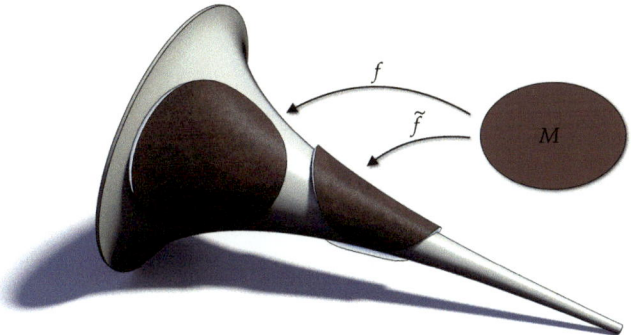

Fig. 6.7 A leather patch fitted to the pseudosphere (at first in one place) is able to slide around freely and even rotate freely

Suppose that $M \subset \{(u, v) \in \mathbb{R}^2 \mid v > 1\}$ and define

$$f \colon M \to \mathbb{R}^3, \quad f(u, v) = \begin{pmatrix} \frac{\cos(u)}{v} \\ \frac{\sin(u)}{v} \\ \log\left(\sqrt{v^2 - 1} + v\right) - \frac{\sqrt{v^2 - 1}}{v} \end{pmatrix}.$$

Choose $\lambda > 1$ and $\mu \in \mathbb{R}$ and define $\tilde{f} \colon M \to \mathbb{R}^3$ by $\tilde{f}(u, v) = f(\lambda u + \mu, \lambda v)$.
We leave it to the reader to verify

$$\langle f_u, f_v \rangle = \langle \tilde{f}_u, \tilde{f}_v \rangle = 0$$

$$\langle f_u, f_u \rangle = \langle f_v, f_v \rangle = \langle \tilde{f}_u, \tilde{f}_u \rangle = \langle \tilde{f}_v, \tilde{f}_v \rangle = \frac{1}{v^2}$$

and that therefore f and $\tilde{f}$ are isometric (see Fig. 6.7).

Here is another example, which will also be of interest later when we study minimal surfaces: Suppose that $M \subset \{(u, v) \in \mathbb{R}^2 \mid u > 0\}$. Let $k, \ell \in \mathbb{Z}$ be two integers with $k + \ell \neq -1$. Then define the **Enneper surface** $f \colon M \to \mathbb{R}^3$ by

$$f(u, v) = \begin{pmatrix} u^{2k+1} \frac{\cos((2k+1)v)}{2k+1} - u^{2\ell+1} \frac{\cos((2\ell+1)v)}{2\ell+1} \\ u^{2k+1} \frac{\sin((2k+1)v)}{2k+1} + u^{2\ell+1} \frac{\sin((2\ell+1)v)}{2\ell+1} \\ 2u^{k+\ell+1} \frac{\cos((k+\ell+1)v)}{k+\ell+1} \end{pmatrix}.$$

Again, we leave it to the reader to verify that

$$\langle f_u, f_v \rangle = 0$$

$$|f_u(u, v)| = u^{2k} + u^{2\ell}$$

$$|f_v(u, v)| = u^{2k+1} + u^{2\ell+1}$$

and that for arbitrary $\lambda \in \mathbb{R}$ the analogous formulas also hold for

$$\tilde{f} \colon M \to \mathbb{R}^3$$

$$\tilde{f}(u, v) = f(u, v + \lambda).$$

This means that also here f and $\tilde{f}$ are isometric, so a leather patch has at least one degree of freedom to slide on the surface without stretching (see Fig. 6.8).

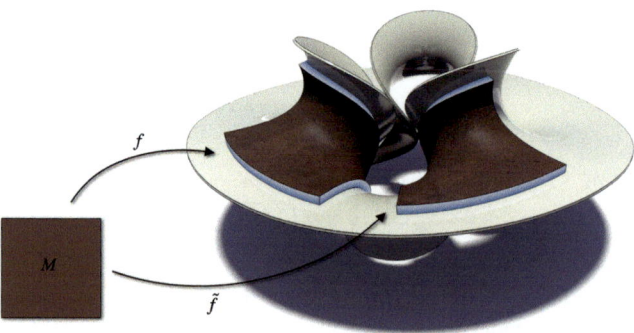

Fig. 6.8 A leather patch fitted to the Enneper surface is able to slide around, but only in one direction. It cannot rotate

Integration and Stokes' Theorem

7

For a curve $\gamma : [a, b] \to \mathbb{R}^n$, global quantities like the length or the bending energy were defined as integrals over arclength of certain functions on $[a, b]$. Before we can define similar quantities for a surface $f : M \to \mathbb{R}^n$, for example, the *area* of f, we have to find a way to integrate functions $g : M \to \mathbb{R}$ in a geometrically meaningful fashion. We will do this in terms of the *area 2-form* det of the metric induced by f. This means that first we have to develop the theory of differential forms on two-dimensional domains M, including the theorem of Stokes.

7.1   Integration on Surfaces

Let $M, \tilde{M} \subset \mathbb{R}^2$ be two compact domains with smooth boundary, $\varphi : \tilde{M} \to M$ an orientation-preserving diffeomorphism and $g : M \to \mathbb{R}$ a smooth function. By the transformation formula for integrals, we have

$$\int_{\tilde{M}} g \circ \varphi \, \det \varphi' = \int_M g.$$

Therefore, if one were to just use $\int_M g$ as the definition for an integral of a function g over the surface f, this integral would not be invariant under reparametrization of f. On the other hand, we are now going to convince ourselves that it is perfectly possible to define the integral of an object like the area form det, as it was introduced in Theorem 6.22:

Definition 7.1

Let M be a compact domain with smooth boundary in $\mathbb{R}^2$. Then a map

$$\sigma : \bigcup_{p \in M} (T_p M \times T_p M) \to \mathbb{R}$$

© The Author(s) 2024

U. Pinkall, O. Gross, *Differential Geometry*, Compact Textbooks in Mathematics, https://doi.org/10.1007/978-3-031-39838-4_7

is called a **2-form** on M if for each $p \in M$ the restriction of σ to $T_p M \times T_p M$ is a skew-symmetric bilinear form and the function

$$\sigma(U, V) \colon M \to \mathbb{R}$$

is smooth.

We denote the set of all 2-forms on M by $\Omega^2(M)$. As a linear subspace of the vector space of all real-valued functions on some set, also $\Omega^2(M)$ is a real vector space. It is also clear how to define the product of a 2-form σ with a function $g \in C^\infty(M)$. One can say that 2-forms are similar to Riemannian metrics, only skew symmetric instead of symmetric and without any non-degeneracy assumptions.

▶ **Remark 7.2** A "model example" of a 2-form which we have already encountered is the well-known $\det \in \Omega^2(M)$.

2-forms are transported under diffeomorphisms by demanding that the transported form applied to the transported tangent vectors yields the same value as before:

Definition 7.3

Let $M, \tilde{M} \subset \mathbb{R}^2$ be two compact domains with smooth boundary, $\varphi \colon \tilde{M} \to M$ a smooth map and σ a 2-form on M. Then we define the **pull-back** of σ under φ as the 2-form $\varphi^*\sigma$ on $\tilde{M}$ that for $p \in \tilde{M}$ and $X, Y \in T_p\tilde{M}$ is given by

$$(\varphi^*\sigma)(X, Y) := \sigma(d\varphi(X), d\varphi(Y)).$$

Theorem 7.4
In the situation of Definition 7.3, the map

$$\Omega^2(M) \to \Omega^2(\tilde{M}), \ \sigma \mapsto \varphi^*\sigma$$

is linear and for $g \in C^\infty(M)$ satisfies

$$\varphi^*(g\sigma) = (g \circ \varphi)(\varphi^*\sigma).$$

The integral over M of a 2-form $\sigma \in \Omega^2(M)$ is defined as follows:

Definition 7.5

The **integral of a 2-form** σ on a compact domain M with smooth boundary in $\mathbb{R}^2$ is defined as

$$\int_M \sigma = \int_M \sigma(U, V)$$

where U and V are the two vector fields on M introduced in Definition 6.10.

The above definition is useful because $\int_M \sigma$ is invariant under pull-back of σ by an orientation-preserving diffeomorphism $\varphi \colon \tilde{M} \to M$.

Theorem 7.6

Let $M, \tilde{M} \subset \mathbb{R}^2$ be two compact domains with smooth boundary, $\varphi \colon \tilde{M} \to M$ an orientation-preserving diffeomorphism and $\sigma \in \Omega^2(M)$ a 2-form. Then

$$\int_{\tilde{M}} \varphi^* \sigma = \int_M \sigma.$$

Proof. Let us write

$$\varphi' = \begin{pmatrix} a & b \\ c & d \end{pmatrix},$$

which means

$$d\varphi(\tilde{U}) = a\, U \circ \varphi + c\, V \circ \varphi$$

$$d\varphi(\tilde{V}) = b\, U \circ \varphi + d\, V \circ \varphi.$$

Therefore, by the skew symmetry of σ and the transformation formula,

$$\int_{\tilde{M}} \varphi^* \sigma = \int_{\tilde{M}} (\varphi^* \sigma)(\tilde{U}, \tilde{V})$$

$$= \int_{\tilde{M}} \sigma(d\varphi(\tilde{U}), d\varphi(\tilde{V}))$$

$$= \int_{\tilde{M}} (ad - bc)\, \sigma(U, V) \circ \varphi$$

$$= \int_M \sigma(U, V)$$

$$= \int_M \sigma.$$

□

In the context of surfaces $f: M \rightarrow \mathbb{R}^n$, we will never integrate functions $g \in C^\infty(M)$ directly, but instead we will first make g into a 2-form by multiplying it with the area form det of the induced metric. Then we can be sure that

$$\int_M g \, \det$$

is a quantity that will stay the same if we reparametrize f as $\tilde{f} = f \circ \varphi$ (and, of course, simultaneously change g to $g \circ \varphi$). Theorem 7.6 above makes it possible to define the area of a Riemannian domain in such a way that it does not change under isometries:

Definition 7.7

The **area of a Riemannian domain** $(M, \langle \, , \rangle)$ is defined as

$$\int_M \det$$

where det is the area form of $\langle \, , \rangle$.

7.2   Integration Over Curves

In order to adequately deal with surfaces, we have found it necessary to add tangent spaces, Riemannian metrics and 2-forms to our toolbox. Let us investigate whether some of these notions might be useful already in the context of curves. For a curve $\gamma: [a, b] \rightarrow \mathbb{R}$, the analog of the domain M of a surface $f: M \rightarrow \mathbb{R}^n$ is the interval $[a, b]$. The tangent bundle of $[a, b]$ is

$$T[a, b] = [a, b] \times \mathbb{R}$$

and the tangent space at $p \in [a, b]$ is $\{p\} \times \mathbb{R}$. The analog of the vector fields U, V on M is the single vector field $X \in \Gamma([a, b])$ defined as

$$X(p) = (p, 1).$$

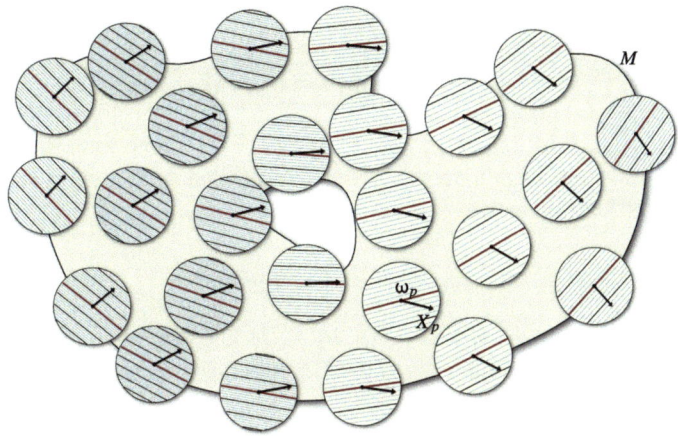

Fig. 7.1 A 1-form $\omega \in \Omega^1(M)$ can be thought of as a smoothly varying ruler which "measures" a vector field $X \in \Gamma(M)$. The spacing of the ruler-lines indicates the "strength" of ω—the closer the spacing, the stronger is ω

The objects that can naturally be integrated over curves are the so-called 1-forms. We will need 1-forms also on planar domains (Fig. 7.1), so we take the opportunity to define also those.

Definition 7.8

Let $[a, b]$ be a closed interval and $M \subset \mathbb{R}^2$ a planar domain with smooth boundary. Smoothness of maps defined on $T[a, b]$ or TM is to be understood in the sense of Definition A.1. Then

(i) A smooth map $\omega\colon T[a, b] \to \mathbb{R}$ is called a **1-form** if its restriction to each tangent space $T_p[a, b]$ is linear. The space of all 1-forms on $[a, b]$ is denoted by $\Omega^1([a, b])$.
(ii) A smooth map $\omega\colon TM \to \mathbb{R}$ is called a **1-form** if its restriction to each tangent space T_pM is linear. The space of all 1-forms on M is denoted by $\Omega^1(M)$.

A general theory of m-forms on domains in $\mathbb{R}^k$ is beyond the scope of this book, so we just collect some special cases that we need:

Definition 7.9

Let $[a, b] \subset \mathbb{R}$ be a closed interval and $M \subset \mathbb{R}^2$ a planar domain with smooth boundary.

(i) If ω is a 1-form on $[a, b]$ and $\varphi\colon [\tilde{a}, \tilde{b}] \to [a, b]$ is a smooth map, then we define the **pull-back** of ω under φ as the 1-form $\varphi^*\omega \in \Omega^1([\tilde{a}, \tilde{b}])$ which is for all $Y \in \Gamma(T[\tilde{a}, \tilde{b}])$ given by

$$(\varphi^* \omega)(Y) = \omega(d\varphi(Y)).$$

(ii) If ω is a 1-form on M and $\varphi \colon \tilde{M} \to M$ is a smooth map, then we define the **pull-back** of ω under φ as the 1-form $\varphi^* \omega \in \Omega^1(\tilde{M})$ which is for all $Y \in \Gamma(T\tilde{M})$ given by

$$(\varphi^* \omega)(Y) = \omega(d\varphi(Y)).$$

(iii) If ω is a 1-form on M and $\gamma \colon [a, b] \to M$ is a smooth map, then we define the **pull-back** of ω under γ as the 1-form $\gamma^* \omega \in \Omega^1([a, b])$ which is for all $Y \in \Gamma(T[a, b])$ given by

$$(\gamma^* \omega)(Y) = \omega(d\gamma(Y)).$$

Definition 7.10

For $\omega \in \Omega^1([a, b])$ we define the **integral of a 1-form** ω over $[a, b]$ as

$$\int_{[a,b]} \omega := \int_a^b \omega(X).$$

Theorem 7.11

Let $\varphi \colon [\tilde{a}, \tilde{b}] \to [a, b]$ be an orientation-preserving diffeomorphism, i.e. a bijective smooth map with $\varphi' > 0$. Then

$$\int_{[\tilde{a},\tilde{b}]} \varphi^* \omega = \int_{[a,b]} \omega.$$

Proof. By the substitution rule and $d\varphi(\tilde{X}) = \varphi' \cdot X \circ \varphi$ we have

$$\int_{[\tilde{a},\tilde{b}]} \varphi^* \omega = \int_{\tilde{a}}^{\tilde{b}} (\varphi^* \omega)(\tilde{X})$$

$$= \int_{\tilde{a}}^{\tilde{b}} \omega(d\varphi(\tilde{X}))$$

$$= \int_{\tilde{a}}^{\tilde{b}} \omega(\varphi' \cdot X \circ \varphi)$$

$$= \int_{\tilde{a}}^{\tilde{b}} \varphi' \cdot \omega(X) \circ \varphi$$

$$= \int_{a}^{b} \omega(X)$$

$$= \int_{[a,b]} \omega.$$

$\square$

Theorem 7.12
Let $M \subset \mathbb{R}^2$ be a compact domain with smooth boundary and $\omega \in \Omega^1(M)$ a 1-form. Let $\tilde{\gamma} : [\tilde{a}, \tilde{b}] \to M$ be a reparametrization of a smooth map $\gamma : [a, b] \to M$, so $\tilde{\gamma} = \gamma \circ \varphi$ for an orientation-preserving diffeomorphism $\varphi : [\tilde{a}, \tilde{b}] \to [a, b]$. Then

$$\int_{[\tilde{a}, \tilde{b}]} \tilde{\gamma}^* \omega = \int_{[a, b]} \gamma^* \omega.$$

Proof. By the chain rule, we have $d\tilde{\gamma} = d\gamma \circ d\varphi$ and therefore

$$\tilde{\gamma}^* \omega = \varphi^*(\gamma^* \omega).$$

Therefore, Theorem 7.11 gives us

$$\int_{[\tilde{a}, \tilde{b}]} \tilde{\gamma}^* \omega = \int_{[\tilde{a}, \tilde{b}]} \varphi^*(\gamma^* \omega) = \int_{[a, b]} \gamma^* \omega.$$

$\square$

As a consequence, we can define the integral of a 1-form ω on M over a curve in M in a way that is invariant under reparametrization:

Definition 7.13

Let $M \subset \mathbb{R}^2$ be a compact domain with smooth boundary and $\omega \in \Omega^1(M)$. Let $\gamma : [a, b] \to M$ be a curve. Then we define

$$\int_{\gamma} \omega := \int_{[a, b]} \gamma^* \omega.$$

In the context of a regular curve $\gamma : [a, b] \to \mathbb{R}^n$, what is the analog of the area form det of a surface $f : M \to \mathbb{R}^n$?

Definition 7.14

The **arclength 1-form** $ds \in \Omega^1([a, b])$ of a curve $\gamma : [a, b] \to \mathbb{R}^n$ is defined as

$$ds(X) = |d\gamma(X)| = |\gamma'|.$$

If we define the arclength function s as in Definition 1.13, the arclength 1-form ds is indeed the derivative of s, which explains the notation. Moreover, if we interpret the left-hand side according to Definition 7.13 and the right-hand side according to Definition 1.14, for a function $g : [a, b] \to \mathbb{R}$ we have

$$\int_{[a,b]} g\, ds = \int_a^b g\, ds.$$

7.3   Stokes' Theorem

When dealing with curves, we frequently used the fundamental theorem of calculus, for example in the form of integration by parts. Also in surface theory we would no get very far without the surface analog of this theorem, which is the so-called Stokes' theorem.

Let $M \subset \mathbb{R}^2$ be a compact connected domain with smooth boundary. The boundary ∂M of M can be parametrized by a finite collection of n closed curves

$$\gamma_j : [a_j, b_j] \to \mathbb{R}^2$$

where $j \in \{1, \ldots n\}$. We assume that each γ_j is oriented in such a way that for any vector $Y \in \mathbb{R}^2$ which at $\gamma_j(x)$ points out of M, we have

$$\det(Y, \gamma_j'(x)) > 0.$$

Given a 1-form $\omega \in \Omega^1(M)$, we make use of Definition 7.13 in order to define the integral of ω over ∂M:

$$\int_{\partial M} \omega = \sum_{j=1}^n \int_{\gamma_j} \omega.$$

> **Theorem 7.15 (Stokes Theorem)**
> Let $M \subset \mathbb{R}^2$ be a compact domain with smooth boundary and $\omega \in \Omega^1(M)$. Then there is a unique 2-form $d\omega \in \Omega^2(M)$ such that for all subdomains $\tilde{M} \subset M$ we have
>
> $$\int_{\tilde{M}} d\omega = \int_{\partial \tilde{M}} \omega.$$
>
> In fact, $d\omega$ is the unique 2-form on M that satisfies
>
> $$d\omega(U, V) = \omega(V)_u - \omega(U)_v.$$

Proof. $\tilde{M}$ could be an arbitrarily small disk around an arbitrary point p in the interior of M, so there can be at most one 2-form σ with the property that for all subdomains $\tilde{M}$

$$\int_{\tilde{M}} \sigma = \int_{\partial \tilde{M}} \omega.$$

This proves the uniqueness part of the claim. If we write

$$\gamma_j' = \begin{pmatrix} \alpha_j \\ \beta_j \end{pmatrix}$$

we have

$$d\gamma_j(X) = \alpha_j \, U \circ \gamma_j + \beta_j \, V \circ \gamma_j$$

and therefore

$$\omega(d\gamma_j(X)) = \alpha_j \, \omega(U) \circ \gamma_j + \beta_j \, \omega(V) \circ \gamma_j.$$

Let us define $\sigma \in \Omega^2(M)$ as the unique 2-form for which

$$\sigma(U, V) = \omega(V)_u - \omega(U)_v.$$

Now we apply Green's theorem from vector calculus to the map

$$Y \colon M \to \mathbb{R}^2, \ Y = \begin{pmatrix} \omega(U) \\ \omega(V) \end{pmatrix},$$

and obtain

$$
\int_M \sigma = \int_M \sigma(U, V)
$$

$$
= \int_M (\omega(V)_u - \omega(U)_v)
$$

$$
= \sum_{j=1}^n \int_{a_j}^{b_j} (\alpha_j\, \omega(U) \circ \gamma_j + \beta_j\, \omega(V) \circ \gamma_j)
$$

$$
= \sum_{j=1}^n \int_{a_j}^{b_j} \omega(d\gamma_j(X))
$$

$$
= \sum_{j=1}^n \int_{a_j}^{b_j} \gamma_j^*\omega
$$

$$
= \int_{\partial M} \omega.
$$

We can apply this argument also to any subdomain $\tilde{M} \subset M$, which proves the existence part of the claim. $\square$

Theorem 7.16

If $\varphi \colon \tilde{M} \to M$ is a smooth map and $\omega \in \Omega^1(M)$, then

$$
\varphi^*(d\omega) = d(\varphi^*\omega).
$$

Proof. The proof is easy if φ is an orientation-preserving diffeomorphism: If $\hat{M} \subset \tilde{M}$ is any subdomain, then by Theorems 7.6, 7.11 and 7.15 we have

$$
\int_{\hat{M}} \varphi^*(d\omega) = \int_{\varphi(\hat{M})} d\omega = \int_{\partial\varphi(\hat{M})} \omega = \int_{\partial\hat{M}} \varphi^*\omega
$$

By the uniqueness part of Theorem 7.15 we then must have $\varphi^*(d\omega) = d(\varphi^*\omega)$.

Unfortunately, here we only assume that φ is a smooth map, so we have to rely on the coordinate formula provided in Theorem 7.15. We use the notation from the proof of Theorem 7.6 together with the equalities $a_{\tilde{v}} = b_{\tilde{u}}$ and $c_{\tilde{v}} = d_{\tilde{u}}$ that follow from the commutativity of partial derivatives of the component functions of φ to compute:

$$d(\varphi^*\omega)(\tilde{U}, \tilde{V})$$

$$= \omega(d\varphi(\tilde{V}))_{\tilde{u}} - \omega(d\varphi(\tilde{U}))_{\tilde{v}}$$

$$= (b \cdot \omega(U) \circ \varphi + d \cdot \omega(V) \circ \varphi)_{\tilde{u}} - (a \cdot \omega(U) \circ \varphi + c \cdot \omega(V) \circ \varphi)_{\tilde{v}}$$

$$= b(a\omega(U)_u \circ \varphi + c\omega(U)_v \circ \varphi) + d(a \cdot \omega(V)_u \circ \varphi + c \cdot \omega(V)_v \circ \varphi$$

$$\quad - a(b \cdot \omega(U)_u \circ \varphi + d \cdot \omega(U)_v \circ \varphi) - c(b \cdot \omega(V)_u \circ \varphi + d \cdot \omega(V)_v \circ \varphi)$$

$$= (ad - bc)(\omega(V)_u - \omega(U)_v) \circ \varphi$$

$$= (ad - bc)\, d\omega(U, V) \circ \varphi$$

$$= d\omega(a \cdot U \circ \varphi + c \cdot V \circ \varphi, b \cdot U \circ \varphi + d \cdot V \circ \varphi)$$

$$= d\omega(d\varphi(\tilde{U}), d\varphi(\tilde{V}))$$

$$= (\varphi^* d\omega)(\tilde{U}, \tilde{V}),$$

which proves the claim. $\qquad\qquad\Box$

Curvature

8

From this chapter on we will focus attention on surfaces $f: M \to \mathbb{R}^3$. The most fundamental tool for analysing such a surface is its unit normal field $N: M \to S^2$ which is a map to the unit sphere $S^2 \subset \mathbb{R}^3$. The derivative of N reveals information about the curvature of f. In particular, the area covered by N on S^2 provides us with a geometric interpretation of the so-called *Gaussian curvature* of f.

8.1 Unit Normal of a Surface in $\mathbb{R}^3$

Most of the material in the Chaps. 6 and 7 was concerned with the intrinsic geometry of Riemannian domains or with surfaces in $\mathbb{R}^n$. From now on we will focus on surfaces $f: M \to \mathbb{R}^3$.

Definition 8.1

Let $M \subset \mathbb{R}^2$ be a domain with smooth boundary and $f: M \to \mathbb{R}^3$ a surface. Then there is a unique smooth map $N: M \to \mathbb{R}^3$ with $\langle N, N \rangle = 1$ such that

(i) For all $p \in M$ and all $X \in T_p M$ we have

$$\langle N(p), df(X) \rangle = 0.$$

(ii) For all $p \in M$ and every positively oriented basis X, Y of $T_p M$ we have

$$\det(N(p), df(X), df(Y)) > 0.$$

N is called the **unit normal** of f (see Fig. 8.1).

U. Pinkall, O. Gross, *Differential Geometry*, Compact Textbooks in Mathematics, https://doi.org/10.1007/978-3-031-39838-4_8

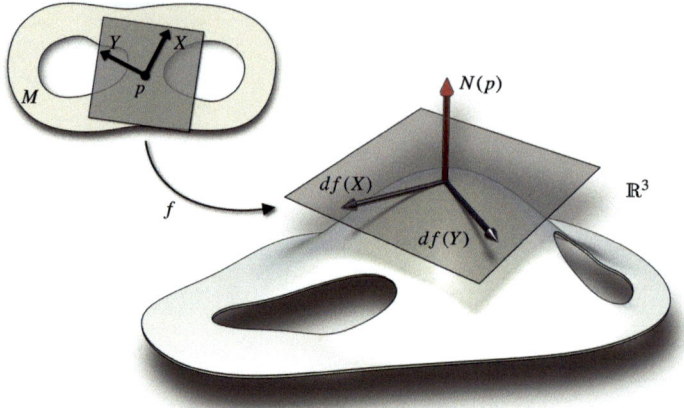

Fig. 8.1 The normal vector $N(p)$ of a surface f at a point p

In terms of the coordinate vector fields U and V we can express N as

$$N = \frac{f_u \times f_v}{|f_u \times f_v|}.$$

Theorem 8.2
For all $p \in M$ and all $X, Y \in T_p M$ we have

$$df(JX) = N(p) \times df(X)$$

$$\det{}_f(X, Y) = \det(N(p), df(X), df(Y)).$$

For the area of f we get

$$\mathcal{A}(f) = \int_M \det{}_f = \int_M \det{}_f(U, V) = \int_M \det(N, f_u, f_v) = \int_M |f_u \times f_v|.$$

Similar as for a surface $f : M \to \mathbb{R}^3$, we can consider the derivative dN of the unit normal $N : M \to \mathbb{R}^3$. In the case of plane curves the derivative of the normal N gave us the curvature κ via the equation

$$N' = \kappa \gamma'.$$

In order to find the analogous equation for surfaces, let us consider a vector field $X \in \Gamma(TM)$ and take the derivative in the direction of X of the equation $1 = \langle N, N \rangle$:

$$0 = d_X \langle N, N \rangle = 2 \langle d_X N, N \rangle.$$

This means that for all $X \in T_p M$ the vector $dN(X)$ lies in the image of the restriction of df to $T_p M$. Therefore, there is a vector $Y \in T_p M$ such that $dN(X) = df(Y)$. Obviously, the dependence of Y on X is linear, so there is a linear map $A_p \colon T_p M \to T_p M$ such that for all $X \in T_p M$ we have

$$dN(X) = df(AX).$$

We leave it to the reader to check that A is a smooth endomorphism field on M.

Definition 8.3

The smooth endomorphism field A is called the **shape operator** of f.

Theorem 8.4

The shape operator A is a self-adjoint endomorphism field with respect to the induced metric, i.e. for all $X, Y \in \Gamma(TM)$ we have

$$\langle AX, Y \rangle = \langle X, AY \rangle.$$

Proof. Since at each point $p \in M$ the two vectors $U(p), V(p)$ form a basis of $T_p M$, it is sufficient to prove the theorem in the special case $X = U, Y = V$. Using the fact that

$$\langle N, df(U) \rangle = \langle N, df(V) \rangle = 0$$

we obtain

$$
\begin{aligned}
\langle AU, V \rangle &= \langle df(AU), df(V) \rangle \\
&= \langle dN(U), df(V) \rangle \\
&= d_U \langle N, df(V) \rangle - \langle N, d_U df(V) \rangle \\
&= -\langle N, f_{vu} \rangle \\
&= -\langle N, f_{uv} \rangle \\
&= d_V \langle N, df(U) \rangle - \langle N, d_V df(U) \rangle
\end{aligned}
$$

$$= \langle dN(V), df(U) \rangle$$
$$= \langle df(AV), df(U) \rangle$$
$$= \langle AV, U \rangle,$$

where we used that the partial derivatives commute, i.e. $f_{uv} = f_{vu}$.   □

8.2   Curvature of a Surface

The shape operator A of a surface $f : M \to \mathbb{R}^3$ captures all the information about how the surface is curved. In fact it measures deviation from being planar:

Theorem 8.5

Let $M \subset \mathbb{R}^2$ be a connected compact domain with smooth boundary and $f : M \to \mathbb{R}^3$ a surface with shape operator A. Then A vanishes identically if and only if there is a plane $E \subset \mathbb{R}^3$ with $f(M) \subset E$.

Proof. If $f(M) \subset E$ with

$$E = \{ \mathbf{p} \in \mathbb{R}^3 \mid \langle \hat{N}, \mathbf{p} \rangle = c \}$$

for some unit vector $\hat{N} \in \mathbb{R}^3$ and $c \in \mathbb{R}$, then

$$\langle \hat{N}, df(X) \rangle = d_X \langle \hat{N}, f \rangle = 0$$

for all $X \in TM$, so the unit normal of f satisfies $N(p) = \pm \hat{N}$ for all $p \in M$. In particular, $dN = 0$ and therefore $A = 0$.

Conversely, by the connectedness of M, $A = 0$ implies that N is constant, i.e. $N(p) = \hat{N}$ for some $\hat{N} \in \mathbb{R}^3$ and all $p \in M$. Then $d\langle \hat{N}, f \rangle = 0$ and (by the connectedness of M) there is $c \in \mathbb{R}$ such that $\langle \hat{N}, f(p) \rangle = c$ for all $p \in M$.   □

At a given point, a surface can be curved by a different amount in different directions. We call a vector $X \in TM$ a **direction** if $\langle X, X \rangle = 1$.

Definition 8.6

For a direction $X \in TM$ we define the **directional curvature** $\kappa(X)$ of f in the direction of X as

$$\kappa(X) := \langle AX, X \rangle.$$

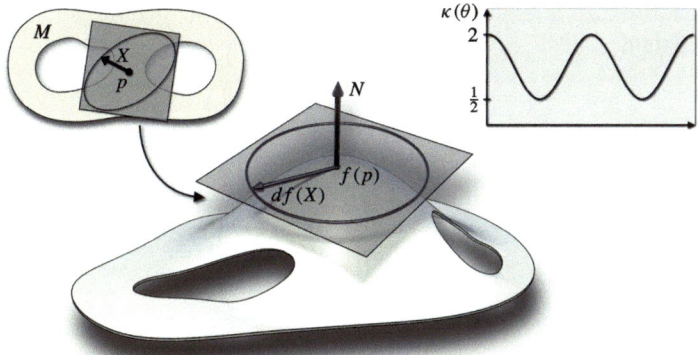

Fig. 8.2 For every unit vector $X \in T_p M$ a surface $f : M \to \mathbb{R}^3$ has a different directional curvature

If X_1, X_2 is an orthonormal basis of $T_p M$ then we can parametrize all unit vectors in $T_p M$ as

$$X(\theta) = \cos\theta \, X_1 + \sin\theta \, X_2.$$

Figure 8.2 contains a plot of the function $\theta \mapsto \kappa(X(\theta))$.

By Theorem 8.4, for all $p \in M$ the linear map

$$A_p := A|_{T_p M} : T_p M \to T_p M$$

is self-adjoint, so there is an orthonormal basis X_1, X_2 in $T_p M$ such that X_1 and X_2 are eigenvectors of A_p:

$$A X_1 = \kappa_1(p) X_1$$
$$A X_2 = \kappa_2(p) X_2.$$

If we assume $\kappa_1 \geq \kappa_2$ the eigenvalue functions $\kappa_1, \kappa_2 : M \to \mathbb{R}$ are well-defined and continuous. They arise from solving the characteristic equation of A_p, in which a square root is involved. This means that in general (if there are points where $\kappa_1(p)$ and $\kappa_2(p)$ coincide) they are not smooth functions.

Definition 8.7

For $p \in M$ the numbers $\kappa_1(p)$ and $\kappa_2(p)$ are called the **principal curvatures** of f at p. A vector $X \in T_p M$ with $\langle X, X \rangle = 1$ is called a principal direction corresponding to the principal curvature κ_j if

$$A X = \kappa_j(p) X.$$

If we parametrize directions $X(\theta)$ at p based on principal directions X_1, X_2 as above we obtain

$$\kappa(\theta) = \langle A(\cos\theta\, X_1 + \sin\theta\, X_2), \cos\theta\, X_1 + \sin\theta\, X_2 \rangle$$
$$= \kappa_1(p)\cos^2\theta + \kappa_2(p)\sin^2\theta$$
$$= \frac{\kappa_1(p) + \kappa_2(p)}{2} + \frac{\kappa_1(p) - \kappa_2(p)}{2}\cos(2\theta).$$

Definition 8.8

The mean value

$$H(p) := \frac{1}{2\pi}\int_0^{2\pi}\kappa(\theta)d\theta$$

is called the **mean curvature** of f at the point p.

We have

$$H(p) = \frac{\kappa_1(p) + \kappa_2(p)}{2} = \frac{1}{2}\mathrm{tr}(A_p),$$

so the function $H: M \to \mathbb{R}$ is smooth.

Definition 8.9

The smooth function

$$K: M \to \mathbb{R},\ \ K(p) = \det A_p = \kappa_1(p)\kappa_2(p)$$

is called the **Gaussian curvature** of f.

If $K(p) > 0$ then the directional curvatures at p are either all positive or all negative. In the first case, the surface looks convex when viewed from "outside" (when we think of N as pointing "outward"). Otherwise it looks concave. Figure 8.3 shows surfaces whose Gaussian curvature is positive everywhere on M.

If $K(p) < 0$ Then the surface bends towards $N(p)$ is some directions and away from N in other directions. Figure 8.4 shows surfaces whose Gaussian curvature is negative everywhere on M.

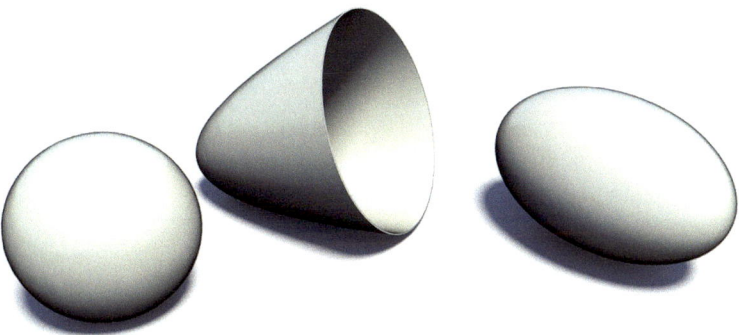

Fig. 8.3 Three surfaces with positive Gaussian curvature

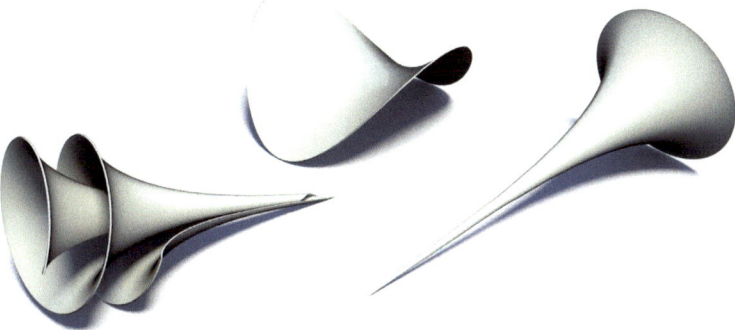

Fig. 8.4 Three surfaces with negative Gaussian curvature

Points where the principal curvatures coincide (and therefore all directions are principal directions) are special and we give them a name:

Definition 8.10

A point $p \in M$ is called an **umbilic point** of the surface f if at p the surface has the same curvature in all directions, i.e. for all directions $X \in T_p M$ we have

$$\kappa(X) = H(p).$$

The most interesting theorems in Differential Geometry lead from local assumptions (curvature properties at each given point) to conclusions about global shape. Here is our first theorem of this kind in the context of surfaces:

Definition 8.11

A subset $S \subset \mathbb{R}^3$ of the form

$$S = \{\mathbf{p} \in \mathbb{R}^3 \mid \langle \mathbf{p} - \mathbf{m}, \mathbf{p} - \mathbf{m} \rangle = r^2\}$$

with $\mathbf{m} \in \mathbb{R}^3$ and $r > 0$ is called a **round sphere**.

Theorem 8.12 (Umbilic Point Theorem)

Let $M \subset \mathbb{R}^2$ be a connected compact domain with smooth boundary and $f : M \to \mathbb{R}^3$ a surface. Then the following are equivalent:

(i) All points $p \in M$ are umbilic points.
(ii) Either $f(M) \subset E$ for some plane $E \subset \mathbb{R}^3$ or $f(M) \subset S$ for some round sphere

$$S = \left\{ \mathbf{p} \in \mathbb{R}^3 \mid \langle \mathbf{p} - \mathbf{m}, \mathbf{p} - \mathbf{m} \rangle = r^2 \right\}.$$

with center $\mathbf{m}$ and radius $r > 0$.

Proof. If $f(M)$ is contained in a plane, we already know that $A = 0$ and therefore all points are umbilic points. If $f(M)$ is contained in a round sphere, then there is a point $\mathbf{m} \in \mathbb{R}^3$ and a radius $r > 0$ such that

$$\langle f - \mathbf{m}, f - \mathbf{m} \rangle = r^2.$$

Clearly then, $f - \mathbf{m} \neq 0$ for all $p \in M$. Differentiating the above equation reveals that for all $p \in M$ and all $X \in T_p M$ we have

$$\langle df(X), f - \mathbf{m} \rangle = 0.$$

Therefore, at each $p \in M$ the unit normal of f must be given by

$$N(p) = \pm \frac{1}{r}(f(p) - \mathbf{m}).$$

By the connectedness of M this implies

$$N = \pm \frac{1}{r}(f - \mathbf{m})$$

and therefore all points are umbilic points:

$$dN = \pm \frac{1}{r} df.$$

Conversely, assume that all points $p \in M$ are umbilic points of f. Then

$$H_v f_u + H f_{uv} = N_{uv} = N_{vu} = H_u f_v + H f_{vu}$$

and therefore $H_v f_u - H_u f_v = 0$. By the connectedness of M, this means that H is constant. In the case $H = 0$ we have $A = 0$ and by Theorem 8.5 we know that $f(M)$ is contained in a plane. Otherwise, there is a constant $r > 0$ such that

$$H = \pm \frac{1}{r}.$$

The function

$$\mathbf{m} : M \to \mathbb{R}^3, \quad \mathbf{m}(p) = f(p) \pm r N(p)$$

then satifies $d\mathbf{m} = 0$ and, by the connectedness of M, must be constant. This means that $f(M)$ lies on a sphere around $\mathbf{m}$ with radius r. $\qquad \square$

8.3 Area of Maps Into the Plane or the Sphere

Recall the second formula from Theorem 8.2: The area form det of a surface $f : M \to \mathbb{R}^3$ with unit normal N is given on $X, Y \in T_p M$ by

$$\det{}_f(X, Y) = \det(N(p), df(X), df(Y)).$$

There are situations where we know what N should be, even if f is not a surface but just a smooth map whose derivative $d_p f : T_p M \to \mathbb{R}^3$ might fail to have a two-dimensional image for some $p \in M$: Define the **Euclidean plane** E^2 as the subset of $\mathbb{R}^3$ where the third component is zero. Then at any point $\mathbf{p} \in E^2$ we consider the third basis vector $\mathbf{e}_3$ as the unit normal vector of E^2 at $\mathbf{p}$. Define the unit two-sphere S^2 as the set of all $\mathbf{p} \in \mathbb{R}^3$ with $|\mathbf{p}| = 1$. Then at any point $\mathbf{p} \in S^2$ we consider $\mathbf{p}$ itself as the unit normal vector of S^2 at $\mathbf{p}$.

Definition 8.13

Let $M \subset \mathbb{R}^2$ be a compact domain with smooth boundary and g a smooth map defined on M with values in either E^2 or S^2. Then we define the **covered area**

form $\sigma_g \in \Omega^2(M)$ on $X, Y \in T_pM$ as follows:

(i) For a smooth map $g: M \to E^2$ we define

$$\sigma_g(X, Y) = \det(\mathbf{e}_3, dg(X), dg(Y)).$$

(ii) For a smooth map $g: M \to S^2$ we define

$$\sigma_g(X, Y) = \det(g(p), dg(X), dg(Y)).$$

If we identify E^2 with $\mathbb{R}^2$ in the obvious way and use the standard determinant det on $\mathbb{R}^2$, the first part of the above definition becomes:

Definition 8.14

Let $M \subset \mathbb{R}^2$ be a compact domain with smooth boundary and $g: M \to \mathbb{R}^2$ a smooth map. Then we define the **covered area form** $\sigma_g \in \Omega^2(M)$ on $X, Y \in T_pM$ as

$$\sigma_g(X, Y) = \det(dg(X), dg(Y)).$$

Definition 8.15

Let $M \subset \mathbb{R}^2$ be a compact domain with smooth boundary and g a smooth map defined on M with values in either E^2, S^2 or $\mathbb{R}^2$. Then we define the **area covered by a map** g as

$$\int_M \sigma_g.$$

Given a smooth map $g: M \to \mathbb{R}^2$ from the unit disk M into $\mathbb{R}^2$, we obtain a loop $\gamma: [0, 2\pi] \to \mathbb{R}^2$ defined by

$$\gamma(x) = g\left(\begin{pmatrix} \cos x \\ \sin x \end{pmatrix}\right).$$

Conversely, every loop $\gamma: [0, 2\pi] \to \mathbb{R}^2$ arises in this way:

Theorem 8.16

Let $\gamma: \mathbb{R} \to \mathbb{R}^2$ be a loop and $M \subset \mathbb{R}^2$ the unit disk. Then there is a smooth map $g: M \to \mathbb{R}^2$ such that the Fig. 8.5 becomes a commutative diagram, i.e. $\gamma = g \circ s$.

Fig. 8.5 By Theorem 8.17, the area covered by the map g equals the sector area of the boundary loop γ, i.e. the blue region is counted positively, the orange region is counted negatively and the region with mixed color is not counted at all

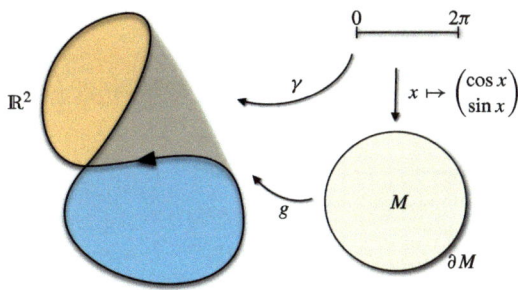

Proof. Let $\varphi \colon [0, 1] \to \mathbb{R}$ be a smooth function such that

$$\varphi(1) = 1$$

$$\varphi(x) = 0 \qquad \text{for} \quad x < \frac{1}{3}.$$

Then we can define $g \colon M \to \mathbb{R}^2$ as the unique map such that for all $r \in [0, 1]$ and all $t \in \mathbb{R}$ we have

$$g\left(r\begin{pmatrix} \cos t \\ \sin t \end{pmatrix}\right) = \varphi(r)\gamma(t).$$

$\square$

Theorem 8.17

Let $\gamma \colon [0, 2\pi] \to \mathbb{R}^2$ be a loop, $M \subset \mathbb{R}^2$ the unit disk and $g \colon M \to \mathbb{R}^2$ any smooth map such that Fig. 8.5 is a commutative diagram. Then the area covered by g equals the sector area of γ.

Proof. Define a 1-form $\omega \in \Omega^1(M)$ by setting for $X \in T_p M$

$$\omega(X) = \frac{1}{2} \det(g(p), dg(X)).$$

Then

$$
\begin{aligned}
2d\omega(U, V) &= d_U \det(g, dg(V)) - d_V \det(g, dg(U)) \\
&= \det(d_U g, d_V g) + \det(g, d_U d_V g) \\
&\quad - \det(d_V g, d_U g) - \det(g, d_V d_U g) \\
&= 2\sigma_g(U, V).
\end{aligned}
$$

Our claim now follows from Stokes Theorem. With

$$\tilde{\gamma}(t) := \begin{pmatrix} \cos t \\ \sin t \end{pmatrix}$$

we obtain

$$
\frac{1}{2} \int_0^{2\pi} \det(\gamma, \gamma') = \frac{1}{2} \int_0^{2\pi} \det(g \circ \tilde{\gamma}, (g' \circ \tilde{\gamma})\tilde{\gamma}')
$$

$$
= \int_{[0,2\pi]} \tilde{\gamma}^* \omega
$$

$$
= \int_{\partial M} \omega
$$

$$
= \int_M d\omega
$$

$$
= \int_M \sigma_g.
$$

□

By Definition 8.13, we obtain a similar interpretation for the area covered by a map $g : M \to S^2$. For us, the most important case is $g = N$ where N is the unit normal of a surface $f : M \to \mathbb{R}^3$ (see Fig. 8.6):

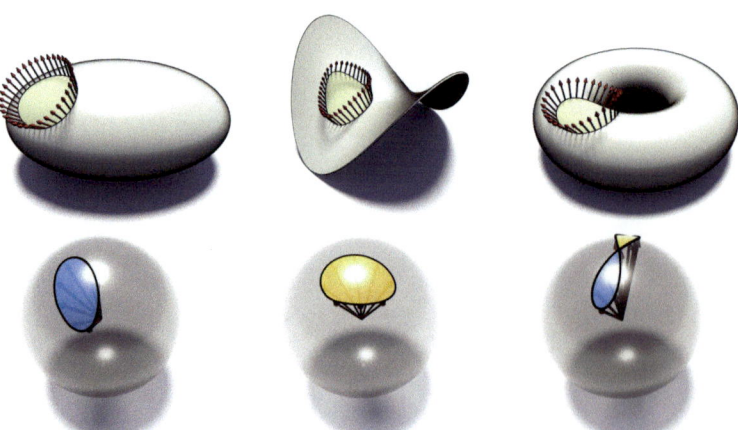

Fig. 8.6 If the Gaussian curvature K is positive in some subregion $\tilde{M} \subset M$, the normal map N will be orientation-preserving in $\tilde{M}$ *(left)*. If K is negative, N will be orientation-reversing *(middle)*. On the right we see a situation where K changes sign in $\tilde{M}$

Theorem 8.18
Let $f : M \to \mathbb{R}^3$ be a surface with unit normal N and Gaussian curvature K. Then the covered area form of N is

$$\sigma_N = K \det{}_f.$$

Proof. For vector fields $X, Y \in \Gamma(TM)$ we have

$$\sigma_N(X, Y) = \det(N, dN(X), dN(Y))$$
$$= \det(N, df(AX), df(AY))$$
$$= \det{}_f(AX, AY)$$
$$= \det A \det{}_f(X, Y) = K \det{}_f(X, Y).$$

$\square$

Levi-Civita Connection

9

The *Levi-Civita connection* of a surface $f: M \to \mathbb{R}^3$ provides a geometrically meaningful way to take directional derivatives of a vector field Y on M. Based on the Levi-Civita connection we derive two important equations that are satisfied by the curvature of a surface in $\mathbb{R}^3$: the *Gauss equation* and the *Codazzi equation*.

9.1 Derivatives of Vector Fields

Let $f: M \to \mathbb{R}^3$ be a regular surface in $\mathbb{R}^3$ with unit normal $N: M \to \mathbb{R}^3$. Let $Y \in \Gamma(TM)$ be a smooth vector field on M, $p \in M$ and $X \in T_pM$. Then, differentiating $0 = \langle N, df(Y) \rangle$ in the direction of X, we obtain

$$0 = d_X \langle N, df(Y) \rangle = \langle df(AX), df(Y) \rangle + \langle N, d_X df(Y) \rangle.$$

Fixing a point $p \in M$, we can decompose every vector $\mathbf{v} \in \mathbb{R}^n$ uniquely as

$$\mathbf{v} = \lambda N + df(Z)$$

for some $\lambda \in \mathbb{R}$ and some $Z \in T_pM$. The vector λN is called the **normal part** of $\mathbf{v}$ and $df(Z)$ is called the **tangential part** of $\mathbf{v}$. In our case, the normal part of $d_X df(Y)$ equals $-\langle AX, Y \rangle N$. The tangential part is of the form $df(Z)$ for some vector $Z \in T_pM$ that we denote by $(\nabla Y)(X)$ or also by $\nabla_X Y$. This gives us

$$d_X df(Y) = -\langle AX, Y \rangle N + df(\nabla_X Y).$$

We leave it to the reader to show that the map $Y \mapsto \nabla Y$ is linear.

© The Author(s) 2024

U. Pinkall, O. Gross, *Differential Geometry*, Compact Textbooks in Mathematics, https://doi.org/10.1007/978-3-031-39838-4_9

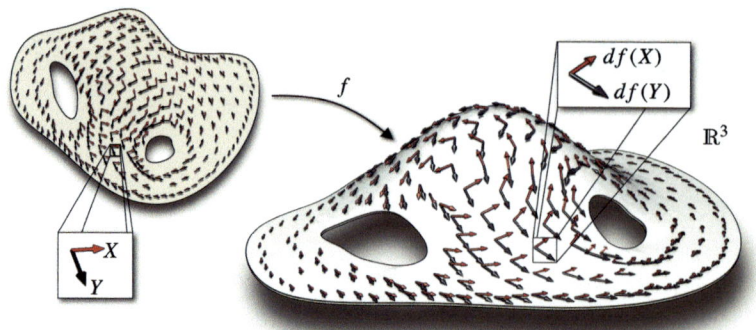

Fig. 9.1 Two vector fields $X, Y \in \Gamma(TM)$

Definition 9.1

The linear map $\nabla \colon \Gamma(TM) \to \Gamma(\mathrm{End}\, TM)$ that assigns to a vector field Y the endomorphism field ∇Y is called the **Levi-Civita connection** of f.

$\nabla_X Y$ can be interpreted as the directional derivative of the vector field Y in the direction of X (Fig. 9.1).

Here are some useful properties of the Levi-Civita connection:

Theorem 9.2
Let X, Y, Z be vector fields on M and $\lambda \colon M \to \mathbb{R}$ a smooth function. Then

(i) $\nabla_X (\lambda Y) = (d_X \lambda) Y + \lambda \nabla_X Y$
(ii) $d_X \langle Y, Z \rangle = \langle \nabla_X Y, Z \rangle + \langle Y, \nabla_X Z \rangle$

Proof. Equation (i) is left as an exercise. For equation (ii) we have

$$
\begin{aligned}
d_X \langle Y, Z \rangle &= d_X \langle df(Y), df(Z) \rangle \\
&= \langle d_X df(Y), df(Z) \rangle + \langle df(Y), d_X df(Z) \rangle \\
&= \langle df(\nabla_X Y), df(Z) \rangle + \langle df(Y), df(\nabla_X Z) \rangle \\
&= \langle \nabla_X Y, Z \rangle + \langle Y, \nabla_X Z \rangle.
\end{aligned}
$$

$\square$

We can also use ∇ to define directional derivatives of endomorphism fields (as defined in Definition 6.19):

Theorem 9.3

If B is a smooth endomorphism field on M and $X \in \Gamma(TM)$ is a vector field, then there is a unique smooth endomorphism field $(\nabla_X B)$ on M such that for all $Y \in \Gamma(TM)$ the following Leibniz rule holds:

$$\nabla_X(BY) = (\nabla_X B)Y + B(\nabla_X Y).$$

Proof. Define C as the unique endomorphism field on M for which

$$CU = \nabla_X(BU) - B\nabla_X U$$
$$CV = \nabla_X(BV) - B\nabla_X V.$$

Clearly, if the endomorphism field $\nabla_X B$ exists, it has to be equal to C. This proves the uniqueness part of the theorem. To prove existence, we show that C has the property we claim for $\nabla_X B$. Let us write Y as a linear combination of U and V:

$$Y = aU + bV.$$

Then, by Theorem 9.2,

$$\begin{aligned}
\nabla_X(BY) &= \nabla_X(aBU + bBV) \\
&= (d_X a)BU + a\nabla_X(BU) + (d_X b)BV + b\nabla_X(BV) \\
&= (d_X a)BU + a(CU + B\nabla_X U) + (d_X b)BV + b(CV + B\nabla_X V) \\
&= CY + B\nabla_X Y.
\end{aligned}$$

$\square$

A smooth **endomorphism field** B for which $\nabla B = 0$ is said to be **parallel**.

Theorem 9.4

Let X, Y, Z be vector fields on M and $\lambda \colon M \to \mathbb{R}$ a smooth function. Then

(i) $\nabla_X(JY) = J\nabla_X Y$
(ii) $d_X \det(Y, Z) = \det(\nabla_X Y, Z) + \det(Y, \nabla_X Z).$

Proof. For Equation (i) we choose a positively oriented orthonormal vector fields $X, Y \in \Gamma(TM)$. Then $JX = Y$ and $JY = -X$ and any vector field $W \in \Gamma(TM)$ can be written as $W = \beta X + \delta Y$. Then for $Z \in \Gamma(TM)$

$$\nabla_Z(JW) = \nabla_Z(\beta Y - \delta X) = (d_Z\beta)Y + \beta\nabla_Z Y - (d_Z\delta)X - \delta\nabla_Z X$$

where we use equation (ii) of Theorem 9.2 for the last equality. Moreover,

$$J\nabla_Z W = J\left((d_Z\beta)X + \beta\nabla_Z X + (d_Z\delta)Y + \delta\nabla_Z Y\right)$$
$$= (d_Z\beta)Y + \beta J\nabla_Z X - (d_Z\delta)X + \delta J\nabla_Z Y .$$

We define $\omega(Z) := \langle \nabla_Z X, Y \rangle$ and conclude that

$$\nabla_Z(JW) = (d_Z\beta)Y - \beta\,\omega(Z)X - (d_Z\delta)X - \delta\,\omega(Z)Y = J\nabla_Z W .$$

In physics, the 1-form ω is known as the **angular velocity** of the orthonormal basis $X_p, Y_p \in T_p M$ in the direction $Z_p \in T_p M$.

Now equation (ii) follows:

$$d_X \det(Y, Z) = d_X \langle JY, Z \rangle$$
$$= \langle \nabla_X(JY), Z \rangle + \langle JY, \nabla_X Z \rangle$$
$$= \langle J(\nabla_X Y), Z \rangle + \langle JY, \nabla_X Z \rangle$$
$$= \det(\nabla_X Y, Z) + \det(Y, \nabla_X Z).$$

□

▶ **Remark 9.5** Note that the equations from 9.2 and the second equation from 9.4 have the flavor of a Leibniz rule.

Theorem 9.6
For the coordinate vector fields we have

$$\nabla_U V = \nabla_V U.$$

Proof. This can be seen by looking at the tangential component of

$$- \langle AU, V \rangle N + df(\nabla_U V) = f_{vu} = f_{uv} = -\langle AV, U \rangle N + df(\nabla_V U).$$

$\square$

9.2   Equations of Gauss and Codazzi

The derivative ∇A (defined in Theorem 9.3) of the shape operator A of a surface $f : M \to \mathbb{R}$ has an important symmetry property:

> **Theorem 9.7**
> *For all vector fields $X, Y \in \Gamma(TM)$ the **Codazzi equation** holds:*
>
> $$(\nabla_X A)Y = (\nabla_Y A)X.$$

Proof. We can write X and Y as linear combinations (with functions as coefficients) of U and V. If we expand both sides of the equation in question accordingly, we see that it is sufficient to consider the special case $X = U$ and $Y = V$. In this case, our claim follows from the fact that partial derivatives of N commute: Equality for the normal part of

$$
\begin{aligned}
- \langle AU, AV \rangle N + df(\nabla_U (AV)) &= d_U df(AV) \\
&= d_U d_V (N) \\
&= d_V d_U (N) \\
&= d_V df(AU) \\
&= - \langle AV, AU \rangle N + df(\nabla_V (AU))
\end{aligned}
$$

is automatically satisfied, while the tangential part gives us what we want to prove.

$\square$

There is another important relation between the shape operator A and the Levi-Civita connection ∇, the so-called Gauss equation:
If $h : M \to \mathbb{R}^k$ is a smooth function, then the partial derivatives of h commute, i.e.

$$d_U d_V h - d_V d_U h = 0.$$

For vector fields this is not true, and the failure of "partial derivatives" of vector fields to commute is determined by the Gaussian curvature of f:

Theorem 9.8
*For any vector field $Z \in \Gamma(TM)$ the **Gauss equation** holds:*

$$\nabla_U \nabla_V Z - \nabla_V \nabla_U Z = -K \det(U, V) JZ$$

where $K = \det A$ is the Gaussian curvature of f.

Proof. Collecting only the terms that are orthogonal to N in

$$d_U(-\langle AV, Z \rangle N + df(\nabla_V Z)) = d_U d_V df(Z)$$
$$= d_V d_U df(Z)$$
$$= d_V(-\langle AU, Z \rangle N + df(\nabla_U Z))$$

we obtain

$$-\langle AV, Z \rangle AU + \nabla_U \nabla_V Z = -\langle AU, Z \rangle AV + \nabla_V \nabla_U Z.$$

Substituting in Theorem 6.24 AU for X, AV for Y and using

$$\det(AU, AV) = \det A \det(U, V).$$

we arrive at the equality this we wanted to prove.   □

9.3   Theorema Egregium

The following theorem is due to Gauss. He called it the "**Theorema Egregium**", which means "most excellent theorem".

Theorem 9.9 (Theorema Egregium)
Suppose that the surfaces $f, \tilde{f} : M \to \mathbb{R}^3$ induce the same Riemannian metric on M. Then f and $\tilde{f}$ have the same Gaussian curvature $K : M \to \mathbb{R}$.

Proof. By the Gauss equation (Theorem 9.8), it is sufficient to prove that if f and $\tilde{f}$ induce the same Riemannian metric on M, they also induce the same Levi-Civita connection. This in turn follows from Theorem 9.10. □

Theorem 9.10

Suppose that the surfaces $f, \tilde{f} : M \to \mathbb{R}^3$ induce the same Riemannian metric on M. Then the Levi-Civita connections of f and $\tilde{f}$ are identical.

Proof. We show that the Levi-Civita connection ∇ induced on M by f is already completely determined by the induced metric $\langle\,,\,\rangle_f$. By Theorem 9.6 and the second equation of Theorem 9.2

$$\langle \nabla_U U, U \rangle = \frac{1}{2} d_U \langle U, U \rangle$$

$$\langle \nabla_U U, V \rangle = d_U \langle U, V \rangle - \langle U, \nabla_U V \rangle$$

$$= d_U \langle U, V \rangle - \langle U, \nabla_V U \rangle$$

$$= d_U \langle U, V \rangle - \frac{1}{2} d_V \langle U, U \rangle$$

$$\langle \nabla_U V, U \rangle = \langle \nabla_V U, U \rangle$$

$$= \frac{1}{2} d_V \langle U, U \rangle$$

$$\langle \nabla_U V, V \rangle = \frac{1}{2} d_U \langle V, V \rangle.$$

Hence $\nabla_U U$ and $\nabla_U V = \nabla_V U$ are completely determined, as well as (by a similar calculation) $\nabla_V V$. Therefore, ∇U and ∇V are completely determined by the knowledge of $\langle\,,\,\rangle_f$ alone. By the first equation of Theorem 9.2, then also ∇Y is determined for an arbitrary vector field $Y = b_1 U + b_2 V$. □

Figure 9.2 reveals the reason why the leather patch on the smoothed dodecahedron in Fig. 6.6 was stuck.

Fig. 9.2 By the Theorema
Egregium, any isometric
motion of the patch has to
preserve Gaussian curvature
(indicated by color), so the
patch cannot slide

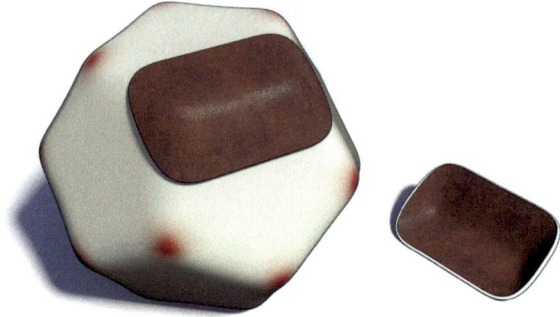

Total Gaussian Curvature $\qquad$ 10

If we know a plane curve $\gamma : [a, b] \rightarrow \mathbb{R}^2$ near its end points, we know its total curvature $\int_a^b \kappa \, ds$ up to an integer multiple of 2π. This follows from the results in Chap. 3. Here we prove a similar result for surfaces $f : M \rightarrow \mathbb{R}^3$ in three-space: If we know f near the boundary of M, we know its total Gaussian curvature $\int_M K \det$ up to an integer multiple of 2π. However, unlike the situation for plane curves, the integer in question is already completely determined by the topology of M.

10.1   Curves on Surfaces

Definition 10.1

Let $f : M \rightarrow \mathbb{R}^3$ be a surface with unit normal field N and $\gamma : [a, b] \rightarrow M$ a curve in M. Then the pair (γ, f) is called a **curve on the surface** f. The space curve

$$\tilde{\gamma} = f \circ \gamma$$

is called the **trace** of (γ, f). The velocity of (γ, f) is defined as $|\tilde{\gamma}'|$ and accordingly the derivative with respect to arclength of a function $g : [a, b] \rightarrow \mathbb{R}^k$ is to be interpreted as

$$\frac{dg}{ds} := \frac{g'}{|\tilde{\gamma}'|}.$$

The unit tangent $\tilde{T}$ of $\tilde{\gamma}$ is called the unit tangent of (γ, f) and the unit normal field

$$\tilde{N} = N \circ \gamma$$

© The Author(s) 2024
U. Pinkall, O. Gross, *Differential Geometry*, Compact Textbooks in Mathematics,
https://doi.org/10.1007/978-3-031-39838-4_10

along $\tilde{\gamma}$ is called the **surface normal** of (γ, f). The unit normal field

$$\tilde{B} = \tilde{T} \times \tilde{N}$$

along $\tilde{\gamma}$ is called the **binormal** of (γ, f).

If (γ, f) is a curve on the surface f, then $(\tilde{\gamma}, \tilde{N})$ defined as above will be a framed curve according to Definition 5.11.

Definition 10.2

If (γ, f) is a curve on the surface f and $\tilde{T}, \tilde{N}, \tilde{B}$ are defined as in Definition 10.1, then

(i) The **normal curvature** of (γ, f) is defined as

$$\kappa_n = \langle \tilde{N}', \tilde{T} \rangle.$$

(ii) The **geodesic curvature** of (γ, f) is defined as

$$\kappa_g = \langle \tilde{B}', \tilde{T} \rangle.$$

(iii) The **geodesic torsion** of (γ, f) is defined as

$$\tau = \langle \tilde{N}', \tilde{B} \rangle.$$

Traditionally, curves on a surface f for which one of these quantities vanishes are designated by special names (see Fig. 10.1):

Definition 10.3

Let (γ, f) be a curve on the surface $f \colon M \to \mathbb{R}^3$. Then

(i) (γ, f) is called an **asymptotic line** if its normal curvature κ_n vanishes.
(ii) (γ, f) is called a **geodesic** if its geodesic curvature κ_g vanishes.
(iii) (γ, f) is called a **curvature line** if its geodesic torsion τ vanishes.

▶ **Remark 10.4** The geodesic in Fig. 10.1 illustrates nicely that geodesics are locally length minimizing, but globally they are not necessarily the shortest path between two points.

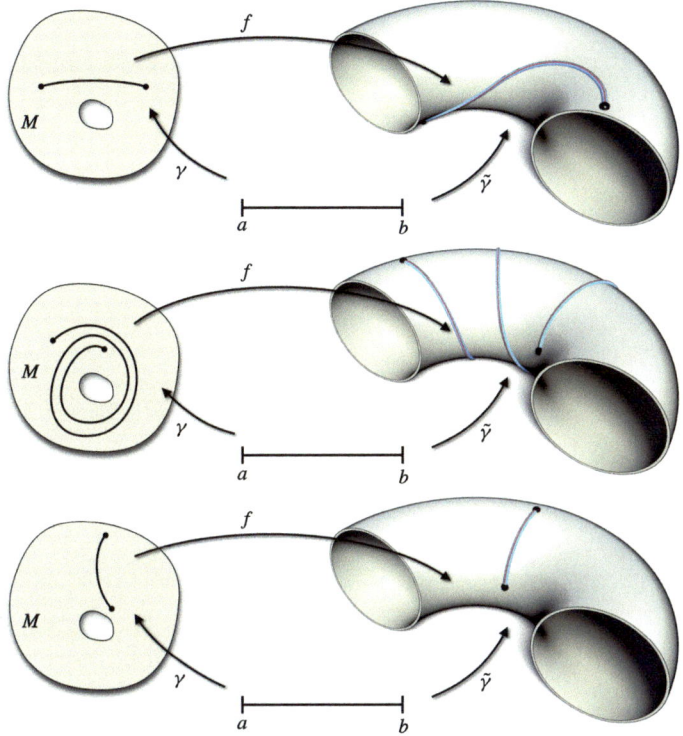

Fig. 10.1 An asymptotic line *(top)*, a geodesic *(middle)* and a curvature line *(bottom)* on a torus

10.2 Theorem of Gauss and Bonnet

Let $M \subset \mathbb{R}^2$ be a compact domain with smooth boundary. By Definition 6.1 and the arguments surrounding Fig. 8.5, each of the n components of the boundary ∂M can be parametrized by a closed curve $\gamma_j \colon [0, 2\pi] \to M$. Given a surface $f \colon M \to \mathbb{R}^3$, we define the **total geodesic curvature** of the boundary ∂M by summing up the integrals of the geodesic curvature κ_g over the corresponding curves (γ_j, f) on the surface f (Definitions 10.1 and 10.2):

$$\int_{\partial M} \kappa_g := \sum_{j=1}^{n} \int_{\gamma_j} \kappa_g \, ds.$$

Definition 10.5

Let $M \subset \mathbb{R}^2$ be a domain with smooth boundary having k components and n boundary curves. Then

$$\chi(M) := 2k - n$$

is called the **Euler characteristic** of M.

Theorem 10.6   (Gauss-Bonnet Theorem)

Let $M \subset \mathbb{R}^2$ be a compact domain with smooth boundary having k connected components. Assume that the boundary ∂M has n components. Let $f : M \to \mathbb{R}^3$ be a surface, $K : M \to \mathbb{R}$ its Gaussian curvature and $\det$ its area form. Then

$$\int_M K \det + \int_{\partial M} \kappa_g \, ds = 2\chi(M).$$

Before we proof the theorem we note that we may always choose a vector field $Z \in \Gamma(TM)$ with $\langle Z, Z \rangle = 1$, for example one could take $Z = \frac{1}{|U|}U$. Moreover, we will make use of the following helpful observations.

Lemma 10.7

Let $M \subset \mathbb{R}^2$ be a compact domain with smooth boundary, $\gamma : [-\pi, \pi] \to \mathbb{R}^2$ be a parametrization of a boundary curve of M and $Z \in \Gamma(TM)$ be a vector field with $\langle Z, Z \rangle = 1$. Then, there is a smooth function $\alpha : \mathbb{R} \to \mathbb{R}$ with $\alpha(x + 2\pi) = \alpha(x) + 2\pi \ell$ for some $\ell \in \mathbb{Z}$ such that

$$T = \cos(\alpha) Z \circ \gamma + \sin(\alpha) JZ \circ \gamma,$$

where $T = \frac{1}{\sqrt{\langle \gamma', \gamma' \rangle}} \gamma'$ is the unit tangent vector of gamma.

Proof. For each $x \in [-\pi, \pi]$ we can choose an $\alpha(x) \in \mathbb{R}$ such that

$$T(x) = \cos(\alpha(x)) Z \circ \gamma(x) + \sin(\alpha(x)) JZ \circ \gamma(x).$$

Due to the smoothness of the boundary curve, locally this choice of α can be made in a smooth fashion. Therefore, $\omega := \alpha'$ is well defined, so that we can safely define

$$\alpha(x) = \alpha(0) + \int_0^x \omega.$$

In particular,

$$\int_{-\pi}^{\pi} \omega = \alpha(\pi) - \alpha(-\pi) = 2\pi\ell$$

for some $\ell \in \mathbb{Z}$, which is exactly the tangent winding number of the curve

$$\begin{pmatrix} \langle \gamma', Z \circ \gamma \rangle \\ \langle \gamma', JZ \circ \gamma \rangle \end{pmatrix} : [-\pi, \pi] \to \mathbb{R}^2$$

which we will denote by $\ell(Z, \langle \cdot, \cdot \rangle)$. $\square$

Lemma 10.8
$\ell(Z, \langle \cdot, \cdot \rangle)$ *is independent of the chosen metric.*

Proof. Let $\langle \cdot, \cdot \rangle^\sim$ be any other metric. Then for any $t \in [0, 1]$ also $\langle \cdot, \cdot \rangle_t :=$ $(1 - t)\langle \cdot, \cdot \rangle + t\langle \cdot, \cdot \rangle^\sim$ is again a Riemannian metric. As ℓ is an integer and depends continuously on t we conclude that it is constant and therefore $\ell(Z, \langle \cdot, \cdot \rangle) = \ell(Z, \langle \cdot, \cdot \rangle^\sim)$. $\square$

Theorem 10.9
Let $M \subset \mathbb{R}^2$ be a compact domain with smooth boundary having k connected components and n boundary curves. Let $\gamma_1, \ldots, \gamma_n$ be closed parametrizations of the n boundary curves of M and $\ell_1, \ldots, \ell_n$ be the tangent winding numbers of $\gamma_1, \ldots, \gamma_n$ with respect to a unit vector field $Z \in \Gamma(TM)$. Then

$$\sum_{j=1}^n \ell(Z) = 2\pi \chi(M).$$

Proof. Without loss of generality let $\langle \cdot, \cdot \rangle = \langle \cdot, \cdot \rangle_{\mathbb{R}^2}$. For $j = 1, \ldots, n$ denote the tangent winding number of γ_j by $\ell_j(Z)$. Then either $\ell_j(Z) = 1$ (if γ_j parametrizes the outer boundary of one of the components of M) or $\ell_j(Z) = -1$. Since there are

k components and $n - k$ interior components, we have

$$\sum_{j=1}^{n} \ell_j(Z) = k - (n - k) = 2k - n = \chi(M).$$

$\square$

Proof of Theorem 10.6 Choose a vector field $Z \in \Gamma(TM)$ with $\langle Z, Z \rangle = 1$ and define a 1-form $\eta \in \Omega^1(M)$ by

$$\eta(X) = \langle \nabla_X Z, J Z \rangle.$$

Think of $\eta(X)$ as the rotation speed of Z in the direction of X. Because of $\langle \nabla_X Z, Z \rangle = 0$ (which follows from differentiating $\langle Z, Z \rangle = 1$) and and since Z, JZ is a positively oriented basis of $T_p M$ we must have

$$\nabla_X Z = \eta(X) J Z.$$

Using this, (ii) of Theorem 9.2, the Gauss equation (Theorem 9.8) and (i) of Theorem 9.4 we find

$$\begin{aligned}
d\eta(U, V) &= d_U \eta(V) - d_V \eta(U) \\
&= \langle \nabla_U \nabla_V Z, J Z \rangle + \langle \nabla_V Z, J \nabla_U Z \rangle \\
&\quad - \langle \nabla_V \nabla_U Z, J Z \rangle - \langle \nabla_U Z, J \nabla_V Z \rangle \\
&= \langle \nabla_U \nabla_V Z, J Z \rangle - \langle \nabla_V \nabla_U Z, J Z \rangle \\
&= -K \det(U, V)
\end{aligned}$$

and therefore

$$d\eta = -K \det.$$

In particular, this means that $d\eta$ does not depend on our choice of Z. Therefore Stokes' theorem implies

$$-\int_M K \det = \int_{\partial M} \eta.$$

For each of the boundary components, the geodesic curvature of $\tilde{\gamma}_j := f \circ \gamma_j$ can be expressed as $\kappa_j := \langle \tilde{T}_j', \tilde{N} \times \tilde{T}_j \rangle$. Then, with

$$(df(Z \circ \gamma))' = -\langle A\gamma', Z \circ \gamma \rangle \tilde{N} + df(\nabla_{\gamma'} Z \circ \gamma)$$

we have

$$\kappa_j = \langle \tilde{T}'_j, \tilde{N} \times \tilde{T} \rangle$$
$$= \langle (df(T_j))', \tilde{N} \times df(T) \rangle$$
$$= \alpha' + \eta(\gamma').$$

Putting everything together we obtain

$$\int_{\partial M} \kappa_g = \sum_{j=1}^{n} \int_{\gamma_j} \kappa_j$$
$$= \sum_{j=1}^{n} \int_{\gamma_j} (\alpha' + \eta)$$
$$= 2\pi \sum_{j=1}^{n} \ell_j(Z) + \int_{\partial M} \eta$$
$$= 2\pi \chi(M) - \int_{M} K \det.$$

$\square$

It is quite striking that the total amount of Gaussian curvature (in the sense of $\int_M K \det$) is completely determined by the geometry of f near the boundary of M (see Fig. 10.2).

Fig. 10.2 Even if we do not know the shape of a rounded cone near its tip (only revealed under a microscope), the integral of the Gaussian curvature can be deduced from the opening angle of the cone

Example 10.10

Suppose M is a disk which is mapped to the top half of a round sphere with radius $r > 0$ by f. Then the boundary curve lies on the equator which is known to be a geodesic, i.e. $\kappa_g = 0$. Therefore, the Gauss-Bonnet theorem yields

$$\int_M K \det = \frac{1}{r^2} \mathcal{A}(f(M)) = 2\pi.$$

Example 10.11

If M is an annulus and f maps it onto a cylinder, then $\chi(M) = 0$ and $K = 0$, so the Gauß-Bonnet formula yields

$$\int_{\partial M} \kappa_g \, ds = 0.$$

10.3   Parallel Transport on Surfaces

In Sect. 5.1 we studied the normal transport $\mathcal{P} \colon T(a)^\perp \to T(b)^\perp$ of a curve $\gamma \colon [a, b] \to \mathbb{R}^3$ with unit tangent T. A closer look reveals that in order to define $\mathcal{P}$ only the smooth map $T \colon [a, b] \to S^2$ is needed. Therefore, given a surface $f \colon M \to \mathbb{R}^3$ and a smooth map $\gamma \colon [a, b] \to M$, we can we can use the same strategy in order to transport tangent vectors $W \in T_{\gamma(a)} M$ to tangent vectors $\mathcal{P}(W) \in T_{\gamma(b)} M$:

Definition 10.12

Let $f \colon M \to \mathbb{R}^3$ be a surface with unit normal field N and $\gamma \colon [a, b] \to M$ a smooth map. Define $\tilde{N} \colon [a, b] \to S^2$ by

$$\tilde{N} := N \circ \gamma$$

and for $W \in T_{\gamma(a)} M$ define the **parallel transport map** $\mathcal{P}_\gamma(W) \in T_{\gamma(b)} M$ in such a way that

$$df(\mathcal{P}(W)) := Z(b)$$

where $Z \colon [a, b] \to \mathbb{R}^3$ solves the initial value problem

$$Z(a) = df(W)$$
$$Z' = -\langle Z, \tilde{N}' \rangle \tilde{N}.$$

$\tilde{N}$ plays exactly the same role here as T did in Sect. 5.1. Hence, for the same reasons as in Sect. 5.1, we have $\langle Z, \tilde{N} \rangle = 0$ and indeed for all $x \in [a, b]$ the vector $Z(x)$ is an element of $df(T_{\gamma(x)}M)$. Furthermore,

$$\mathcal{P}_\gamma : T_{\gamma(a)}M \to T_{\gamma(a)}M$$

is an orientation-preserving orthogonal map with respect to the metrics induced by f on $T_{\gamma(a)}M$ and $T_{\gamma(b)}M$.

The derivative $Z'(x)$ is a multiple of $N(\gamma(x))$, so it has no component in $df(T_{\gamma(x)}M)$. In the spirit of the Sect. 9.1 (Levi-Civita connection), where a derivative $\nabla_X Y$ of a vector field Y was defined in terms of the tangential component of $d_X(df(Y))$, this means that $\mathcal{P}$ can be viewed as parallel transport along γ.

Imagine a pendulum swinging at a point of a surface $f : M \to \mathbb{R}^3$ subject to gravity pointing away form the unit normal of the surface. Suppose we transport the swinging pendulum along a path $f \circ \gamma$ where $\gamma : [a, b] \to M$ is a smooth map and that the plane in which the pendulum swings initially is given as $df(W)$ where $W \in T_{\gamma(a)}M$ is a unit vector with respect to the induced metric. Then Physics tells us that the plane in which the pendulum swings once it arrives at $f(\gamma(b))$ will be given by the unit vector $df(\mathcal{P}(W))$.

In the special case where f parametrizes the surface of the earth and the movement γ corresponds to the rotation of the earth, this effect can be experimentally verified and is known under the name of **Foucault's pendulum** (see Fig. 10.3).

As in Sect. 5.1, if we choose unit vectors (with respect to the induced metric) $W_a \in T_{\gamma(a)}M$ and $W_a \in T_{\gamma(b)}M$, we can measure the parallel transport along γ

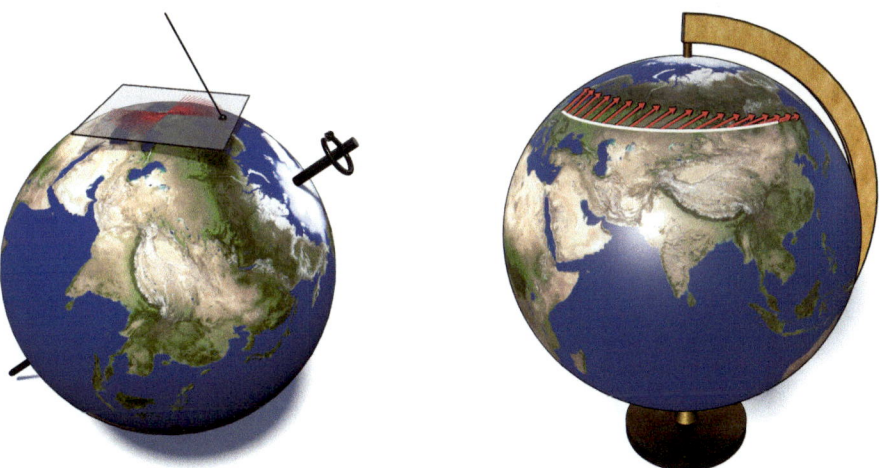

Fig. 10.3 In 1851, Léon Foucault build a *Foucault pendulum* to demonstrate the rotation of the earth *(left)*. The phenomenon can be understood with the concept of parallel transport, where a tangent vector is transported along a circle of latitude *(right)*

Fig. 10.4 Parallel transport of a tangent vector along a closed curve on a surface with positive Gaussian curvature *(left)* and along the boundary of a surface with negative Gaussian curvature *(right)*

by an angle $\mathcal{P}_W \in \mathbb{R}/_{2\pi\mathbb{Z}}$. For closed curves γ this angle does not depend on the choice of W_a and W_b as long as we make sure that $W_a = W_b$. In the special case where γ parametrizes the boundary ∂M of M, this angle can be expressed in terms of the total Gaussian curvature of f (see Fig. 10.4).

> **Theorem 10.13**
> *Suppose that $f \colon M \to \mathbb{R}^3$ is a surface and that M has only a single boundary component parametrized by a curve $\gamma \colon [a, b] \to M$. Then the **monodromy angle** of γ satisfies*
>
> $$\mathcal{M}(\gamma) \equiv \int_M K \det \mod 2\pi\mathbb{Z}$$
>
> *where K is the Gaussian curvature of f.*

Proof. Let us assume that γ has unit speed with respect to the induced metric and therefore $\tilde{T} := \tilde{\gamma}'$ is the unit tangent field of $\tilde{\gamma} := f \circ \gamma$. Define $\tilde{N} := N \circ \gamma$ where N is the unit normal field of f. Let W and Z be defined as in Definition 10.12. Then there is a smooth function $\alpha \colon [a, b] \to \mathbb{R}$ such that

$$Z = \cos\alpha\, \tilde{T} + \sin\alpha\, \tilde{N} \times \tilde{T}.$$

We denote by $\kappa_g = \langle \tilde{T}', \tilde{N} \times \tilde{T} \rangle$ the binormal curvature of the framed curve $(\tilde{\gamma}, \tilde{N})$. Because Z' is normal, we have

$$0 = \langle Z', \tilde{N} \times Z \rangle = \alpha' + \kappa_g.$$

Finally, by the Gauss-Bonnet Theorem 10.6 we have

$$\mathcal{M}(\gamma) \equiv \alpha(b) - \alpha(a) = \int_a^b \alpha' = -\int_a^b \kappa_g \equiv \int_M K \ \det \quad \mathrm{mod}\ 2\pi \mathbb{Z}.$$

$\square$

Closed Surfaces

11

We define a *closed surface* as a surface $f : M \to \mathbb{R}^3$ whose boundary components have been matched in pairs in such a way that f as well as its unit normal N are continuous across the boundary. This allows us to prove an analog of the fact that the tangent winding number of a closed plane curve is an integer: The total Gaussian curvature $\int_M K$ det of a closed surface $f : M \to \mathbb{R}^3$ is equal to $2\pi \chi(M)$ where $\chi(M)$ is the Euler characteristic.

11.1 History of Closed Surfaces

Our goal here is to define "closed surfaces" in such a way that we are able to prove an analog of Theorem 3.8, which says that the turning number of a plane curve is an integer. Furthermore, in Sect. 13.1 we want to discuss for closed surfaces the analog of the total squared curvature of a curve.

Our approach will be based on the very idea that was already at the heart of the 1845 paper by Möbius where closed surfaces were studied for the first time: By cutting them into horizontal slices, Möbius decomposed closed surfaces into pieces each of which can be parametrized by a compact domain with smooth boundary in $\mathbb{R}^2$. Figure 11.1 is adapted from the paper by Möbius. This very idea was already the motivation for us to allow for disconnected domains in the case of surfaces and will be formalized in Sect. 11.2.

More details on the early history of surface theory can be found in an article by Peter Dombrowski [11].

A more advanced way to define closed surfaces in $\mathbb{R}^n$ (that would not need to cut the surface into pieces that can be parametrized by planar domains) would be to define them in terms of smooth maps $f : M \to \mathbb{R}^n$ defined on 2-dimensional compact manifolds M. Such manifolds were first defined in 1910 by Hermann Weyl in a famous book with the title "Die Idee der Riemannschen Fläche" [44].

© The Author(s) 2024
U. Pinkall, O. Gross, *Differential Geometry*, Compact Textbooks in Mathematics,
https://doi.org/10.1007/978-3-031-39838-4_11

Fig. 11.1 Möbius decomposed closed surfaces into pieces that can be parametrized by compact domains in $\mathbb{R}^2$ with smooth boundary (modeled after Möbius' original sketch in [29])

On the other hand, the fully developed version of the Gauss-Bonnet theorem (which we will prove in the next chapter) is already contained in the 1903 thesis of Werner Boy [8], that he did under the supervision of David Hilbert.

Modern treatments of Differential Topology (like the books by Andrew Wallace [42] and Morris Hirsch [16]) often discuss surface topology in their last chapters. The main work there goes into proving (with the help of Morse theory) that indeed every compact 2-dimensional manifold can be decomposed into pieces each of which can be parametrized by a compact domain with smooth boundary in $\mathbb{R}^2$. Therefore, the work that will be done in the next two chapters would not become obsolete even if we had manifolds at our disposal.

11.2 Defining Closed Surfaces

Suppose that for a surface $f : M \to \mathbb{R}^3$ the boundary components of M match up in pairs in such a way that, given suitable parametrizations of the boundary curves, corresponding points of ∂M are mapped to the same points in $\mathbb{R}^3$. If in addition also the unit normals of f fit together up to sign on ∂M, we consider f (together with a specification of the boundary matching) as a closed surface (Fig. 11.2):

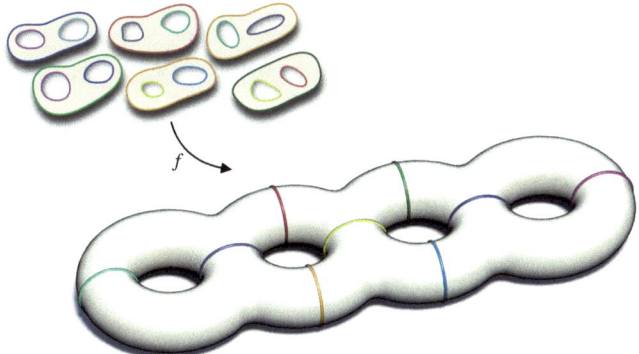

Fig. 11.2 A closed surface f

Let $M \subset \mathbb{R}^2$ be a compact domain with smooth boundary and $f : M \to \mathbb{R}^3$ a surface with unit normal N. We parametrize the boundary curves of M by closed curves

$$\gamma_1, \ldots, \gamma_n : [-\pi, \pi] \to \mathbb{R}^2$$

and define curves $\tilde{\gamma}_1, \ldots, \tilde{\gamma}_n : [-\pi, \pi] \to \mathbb{R}^3$ by

$$\tilde{\gamma}_j := f \circ \gamma_j.$$

As in Definition 10.1, we equip the closed space curves $\tilde{\gamma}_j$ with unit normal fields $\tilde{N}_j := N \circ \gamma_j$. Let

$$\rho : \{1, \ldots, n\} \to \{1, \ldots, n\}$$

a bijective map such that

$$(\rho \circ \rho)(j) = j$$

for all j. Then the pair (f, ρ) is called a **closed surface** if there are signs $\epsilon_1, \ldots, \epsilon_n \in \{-1, 1\}$ such that for all $j \in \{1, \ldots, n\}$ we have:

(i) If $\rho(j) \neq j$ then

$$\tilde{\gamma}_{\rho(j)}(x) = \tilde{\gamma}_j(\epsilon_j x)$$
$$\tilde{N}_{\rho(j)}(x) = -\epsilon_j \tilde{N}_j(\epsilon_j x).$$

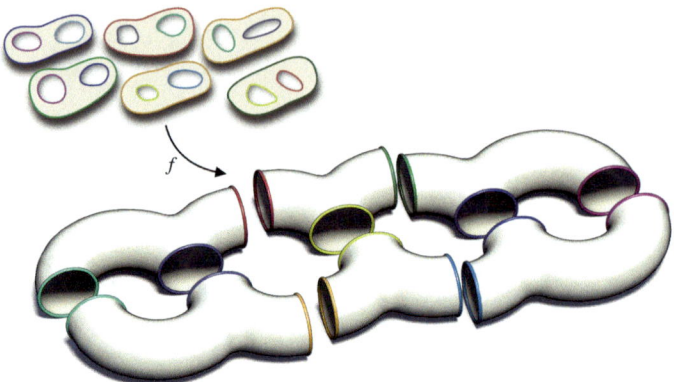

Fig. 11.3 The surface in Fig. 11.2 made into a non-closed surface by applying a small translation to each piece

(ii) If $\rho(j) = j$ then $\epsilon_j = 1$ and

$$\tilde{\gamma}_j(x) = \begin{cases} \tilde{\gamma}_j(x + \pi) & \text{for } x \in [-\pi, 0) \\ \tilde{\gamma}_j(x - \pi) & \text{for } x \in [0, \pi] \end{cases}$$

$$\tilde{N}_j(x) = \begin{cases} -\tilde{N}_j(x + \pi) & \text{for } x \in [-\pi, 0) \\ -\tilde{N}_j(x - \pi) & \text{for } x \in [0, \pi]. \end{cases}$$

It is easy to see that such $\epsilon_1, \dots, \epsilon_n$ are uniquely determined by f and ρ. We say that a closed surface is *oriented* if $\epsilon_j = -1$ for all $j \in \{1, \dots, n\}$.

Figure 11.3 shows the shape of the individual pieces that are being glued in Fig. 11.2. It has $k = 6$ components and $n = 18$ boundary curves.

Here is another example: M now consists of a disk with boundary γ_1 and an annulus with boundary curves γ_2 and γ_3. First, we tentatively define f on the disk bounded by γ_1 and obtain the cap on the upper right of Fig. 11.4. Postponing for the moment the task (indicated by the double-arrow on the right) of gluing γ_1 to γ_2, we first glue γ_3 to itself and obtain a Möbius band (on the bottom of the lower right of Fig. 11.4):

By growing the Möbius band (see Fig. 11.5) we finally obtain the closed surface we wanted to construct:

This surface (fully closed in Fig. 11.6) was found by Werner Boy in 1903 and is called the **Boy surface**.

Figure 11.7 shows two surfaces which are obtained by gluing the boundary curve of an annulus to itself appropriately. Even though both compact domains have $k = 1$ components and $n = 2$ boundary loops, the distinct maps $f, \tilde{f}$ lead to distinct closed surfaces. In particular, although the map ρ is the same, they have opposite sign ϵ.

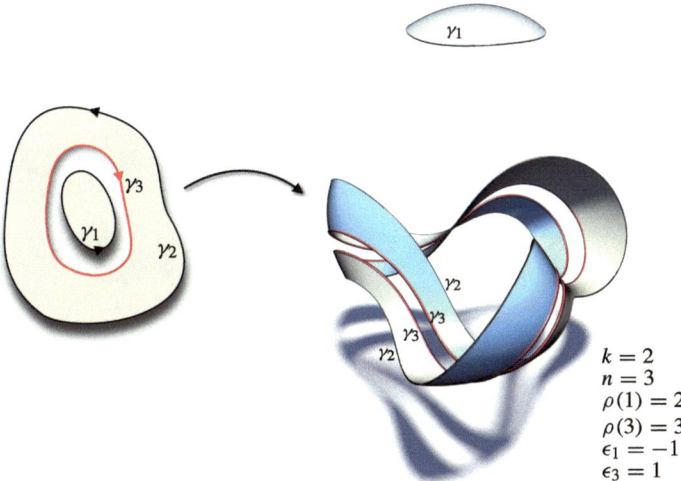

Fig. 11.4 The annulus part of the domain on the left has its red boundary component glued to itself. After growing the resulting Möbius strip, the other boundary component can be glued to the image of the disk part of the domain. The result is the so-called **Boy surface**

11.3 Boy's Theorem

Definition 11.2

We say that a surface $f : M \to \mathbb{R}^3$ **closes up** if there is ρ such that (f, ρ) is a closed surface in the sense of Definition 11.1.

Recall that for every closed plane curve $\gamma : [a, b] \to \mathbb{R}^2$ there was an integer $n \in \mathbb{Z}$ such that

$$\int_a^b \kappa \, ds = 2\pi n.$$

Surprisingly, the analog of this fact in the context of surfaces (cf. Theorem 11.3) does not involve any information about the specific way in which f closes up, but only depends on properties of the domain M. The theorem is a variant of the Gauss-Bonnet Theorem 10.6. Usually, it would be called by the same name. However, historically this is not quite correct. This theorem was in fact the main result of the

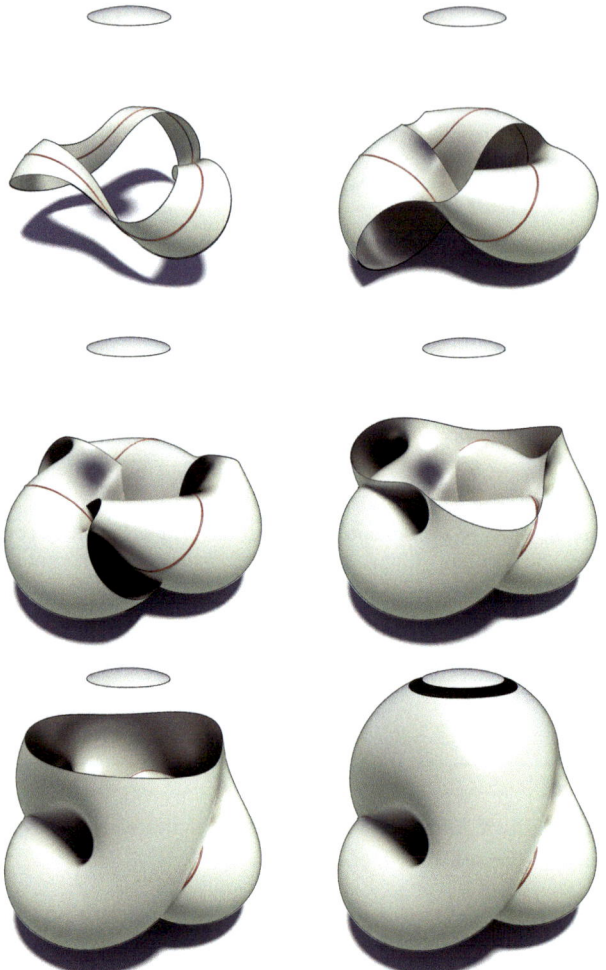

Fig. 11.5 A growing Möbius strip can be capped off to form a Boy surface

thesis of Werner Boy [8], written in 1903 under the supervision of David Hilbert. For this reason, we name it after Boy:

Theorem 11.3 (Boy's Theorem)
Let $f : M \to \mathbb{R}^3$ be a surface that closes up. Then the Gaussian curvature K of f satisfies

$$\int_M K \ \det = 2\pi \ \chi(M).$$

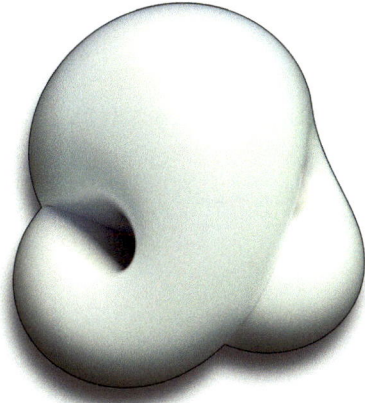

Fig. 11.6 The Boy surface is a closed, non-oriented surface

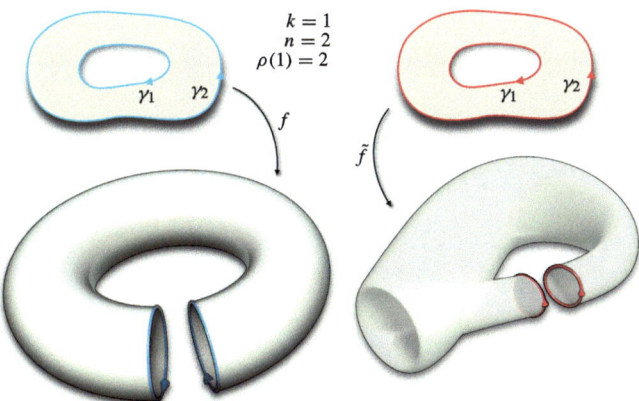

Fig. 11.7 After pushing the two boundary curves together, we obtain a closed surface which is oriented—a torus *(left)*, or a closed surface that is not oriented—a so-called **Klein bottle** *(right)*

Before we give the proof, we introduce the notion of an **orientation cover** of a closed surface. Given a closed surface (f, ρ) with $f : M \to \mathbb{R}^3$, we can define an oriented closed surface $(\tilde{f}, \tilde{\rho})$ in the following way:

Let us use M_{-1} as another name for M and, using an orientation-reversing isometry $g : \mathbb{R}^2 \to \mathbb{R}^2$, we place a second copy $M_1 = g(M)$ into $\mathbb{R}^2$ in such a way that M_{-1} and M_1 are disjoint. Then we define

$$\tilde{M} := M_{-1} \cup M_1$$

and

$$\tilde{f}: \tilde{M} \to \mathbb{R}^3, \quad \tilde{f}(p) = \begin{cases} f(p) & \text{if} \quad p \in M_{-1} \\ (f \circ g^{-1})(p) & \text{if} \quad p \in M_1. \end{cases}$$

We can label the boundary curves of $\tilde{M}$ by the elements of $\{-1, 1\} \times \{1, \ldots, n\}$ and parametrize them by maps

$$\gamma_{(i,j)}: \mathbb{R} \to \partial \tilde{M}, \quad \gamma_{(i,j)} = \begin{cases} \gamma_j & \text{if} \quad i = -1 \\ x \mapsto g \circ \gamma_j(-x) & \text{if} \quad i = 1. \end{cases}$$

Finally, we define

$$\tilde{\rho}: \{-1, 1\} \times \{1, \ldots, n\} \to \{-1, 1\} \times \{1, \ldots, n\}, \quad \tilde{\rho}(i, j) = (-\epsilon_j\, i, \rho(j)).$$

We now leave it to the reader to check that $(\tilde{f}, \tilde{\rho})$ is an oriented closed surface, i.e. we obtain a closed surface by setting $\tilde{\epsilon}_{(i,j)} = -1$ for all $(i, j) \in \{-1, 1\} \times \{1, \ldots, n\}$.

Definition 11.4

The closed surface $(\tilde{f}, \tilde{\rho})$ constructed above is called an **orientation cover** of f.

Proof of Theorem 11.3—Boy's Theorem If ρ has no fixed points (no boundary component is glued to itself), one just has to note that the existence of ρ (making (f, ρ) into a closed surface) implies that in Theorem 10.6 the total geodesic curvatures of the individual boundary curves cancel in pairs. If ρ has fixed points, we note that the $\tilde{\rho}$ of the orientation cover has no fixed points and therefore our theorem holds for $\tilde{f}$. Dividing both sides of the resulting equation by two, we see that our theorem also holds for f.   □

11.4   The Genus of a Closed Surface

The Euler characteristic of a closed surface was solely a property of its domain M, the specific way the various boundary curves are glued is irrelevant for the Euler characteristic. There is another number associated with a closed surface (f, ρ), the so-called **genus**, that depends on the gluing correspondence ρ:

Suppose $M \subset \mathbb{R}^2$ is a domain with k components and n boundary curves. Consider the map that assigns to each $j \in \{1, \ldots, n\}$ the index $c(j) \in \{1, \ldots, k\}$ of the component of M to which the jth boundary component belongs. Let us consider the graph G whose vertex set is $\{1, \ldots, k\}$ and in which two vertices $\ell, \tilde{\ell}$ with $\ell \neq \tilde{\ell}$ are connected by an edge if and only if there is an index $j \in \{1, \ldots, n\}$ for which $c(j) = \ell$ and $c(\rho(j)) = \tilde{\ell}$, which means that the components of M with indices j and $\tilde{j}$ are glued via one (or more) of their respective boundary curves. We say that two vertices ℓ and $\tilde{\ell}$ of G are **connectable** in G if it is possible to travel from ℓ to $\tilde{\ell}$ by following edges. Connectability is an equivalence relation and the corresponding equivalence classes are called the connected components of G.

Definition 11.5

If $\{\ell_1, \ldots, \ell_{\tilde{k}}\}$ is a component of the graph G, then

$$\tilde{f} = f|_{M_{\ell_1} \cup \ldots \cup M_{\ell_{\tilde{k}}}}$$

closes up with boundary gluing $\tilde{\rho}$ read off from (f, ρ). We call the resulting closed surface $(\tilde{f}, \tilde{\rho})$ a **component** of (f, ρ). We call (f, ρ) **connected** if it has only one component.

So the components of a closed surface are in one-to-one correspondence with the components of its associated graph G.

Definition 11.6

Let M be a compact domain with k components and n boundary curves. Let (f, ρ) be a closed surface with $f: M \rightarrow \mathbb{R}^3$. If (f, ρ) has m connected components, we define the **genus** of (f, ρ) as

$$g := \frac{n}{2} - k + m.$$

In terms of the genus, the Gauss-Bonnet formula takes the form

$$\int_M K \, \det = 4\pi (m - g).$$

The first surface featured in Sect. 11.2 has genus $g = 4$, the Klein bottle has genus $g = 1$ und the Boy surface has genus $g = \frac{1}{2}$. The two surfaces in Fig. 11.8 have genus $g = \frac{5}{2}$ and genus $g = 2$ respectively.

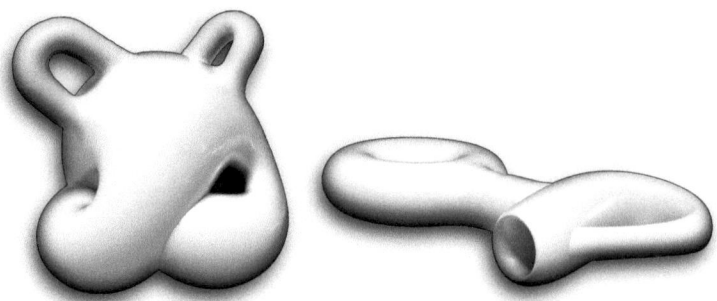

Fig. 11.8 Non-oriented surfaces of genus $g = \frac{5}{2}$ *(left)* and $g = 2$ *(right)*. They are obtained by smoothly gluing handles onto a Boy surface or respectively a Klein bottle

Variations of Surfaces

12

We derive the basics of Vector Calculus on surfaces and explore variations of surfaces. In particular, we compute the variational derivative of the area form det and of the shape operator A. We show that the critical points of the area functional are the surfaces with mean curvature $H = 0$. If we constrain the enclosed volume, the critical points of area are the surface with constant mean curvature. These results mirror the situation for plane curves, where the analogous variational problems lead to straight lines ($\kappa = 0$) or circles ($\kappa = $ const).

12.1 Vector Calculus on Surfaces

Throughout this section, $M \subset \mathbb{R}^2$ is a Riemannian domain, $f: M \to \mathbb{R}^3$ a surface and $\langle\,,\rangle$ its induced metric. We will only use the area form det, the $90°$-rotation J and the Levi-Civita-connection ∇, which by Theorems 6.22, 6.23 and 9.10 are already determined by the induced metric. This means that this section is dealing only with intrinsic geometry.

If $g \in C^\infty(M)$ is a smooth function, then for each $p \in M$ the restriction $(dg)|_{T_pM}$ is a linear map on T_pM and the restriction $\langle\,,\rangle|_{T_pM \times T_pM}$ is a Euclidean scalar product. Therefore, there is a unique vector $Y(p) \in T_pM$ such that $dg(X) = \langle Y(p), X\rangle$ for all $X \in T_pM$. The smoothness of the vector field Y defined in this way follows in the usual way, see for example the proof of Theorem 6.20. This leads us to the following:

Definition 12.1

For $g \in C^\infty(M)$ there is a unique vector field

$$\text{grad } g \in \Gamma(TM)$$

© The Author(s) 2024
U. Pinkall, O. Gross, *Differential Geometry*, Compact Textbooks in Mathematics,
https://doi.org/10.1007/978-3-031-39838-4_12

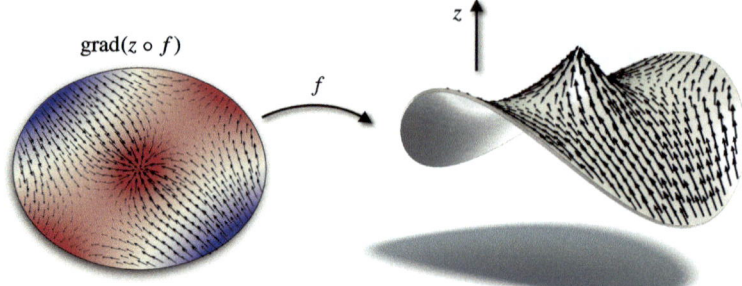

Fig. 12.1 The gradient vector field of the function $z \circ f$ (z being the third coordinate function on $\mathbb{R}^3$) for a surface $f : M \to \mathbb{R}^3$. On the left, the value of $z \circ f$ is indicated by color-coding

characterized by the fact that for all vector fields $X \in \Gamma(TM)$ we have

$$dg(X) = \langle \operatorname{grad} g, X \rangle.$$

The vector field $\operatorname{grad} g \in \gamma(TM)$ is called the **gradient** of g.

So a function $g \in C^\infty(M)$ gives us a vector field $\operatorname{grad} g \in \Gamma(TM)$ (see Fig. 12.1). On the other hand, by taking the trace of the endomorphism field ∇Y, a vector field $Y \in \Gamma(TM)$ gives us a function $\operatorname{div} Y \in C^\infty(M)$.

Definition 12.2

For a vector field $Y \in \Gamma(TM)$ the function

$$\operatorname{div} Y : M \to \mathbb{R}, \ \operatorname{div} Y = \operatorname{tr}(\nabla Y)$$

is called the **divergence** of Y.

The following theorem from Linear Algebra is useful for calculating the trace of an endomorphism field.

Theorem 12.3
Let W be a 2-dimensional vector space with a determinant function $\det$ and $A \colon W \to W$ a linear map. Then for any two vectors $X, Y \in W$ we have

$$\det(AX, Y) - \det(AY, X) = \operatorname{tr} A \, \det(X, Y).$$

Proof. If X and Y are linearly dependent, both sides of the equation vanish. Otherwise, X and Y form a basis of W and we can write

$$AX = aX + cY$$
$$AY = bX + dY.$$

Our claim now follows from

$$\operatorname{tr} A = a + d.$$

$\square$

For the divergence of the product of a function and a vector field we have a Leibniz formula:

Theorem 12.4
For $g \in C^\infty(M)$ and $Z \in \Gamma(TM)$ we have

$$\operatorname{div}(gZ) = \langle \operatorname{grad} g, Z \rangle + g \operatorname{div}(Z).$$

Proof. With the notation $G := \operatorname{grad} g$ and with the help of Theorems 6.24 and 12.3, for $X, Y \in \Gamma(TM)$ we have

$$\begin{aligned}
\operatorname{div}(gZ)\det(X, Y) &= \det(\nabla_X(gZ), Y) - \det(\nabla_Y(gZ), X) \\
&= \det(\langle X, G \rangle Z + g\nabla_X Z, Y) - \det(\langle Y, G \rangle Z + g\nabla_Y Z, X) \\
&= -\det(\langle Y, G \rangle X - \langle X, G \rangle Y, Z) + g \operatorname{div}(Z)\det(X, Y) \\
&= (\det(-JG, Z) + g \operatorname{div}(Z))\det(X, Y) \\
&= (\langle G, Z \rangle + g \operatorname{div}(Z))\det(X, Y).
\end{aligned}$$

$\square$

Definition 12.5

The divergence of the gradient of a function $g \in C^\infty(M)$

$$\Delta g := \operatorname{div} \operatorname{grad} g$$

is called the **Laplacian** of g.

The divergence of a 90°-rotated gradient vanishes:

Theorem 12.6
For every $g \in C^\infty(M)$ we have

$$\text{div}(J\,\text{grad}\,g) = 0.$$

Proof. Using again the notation $G := \text{grad}\,g$, by Theorems 9.6 and 12.3 we obtain

$$
\begin{aligned}
\text{div}(J\,\text{grad}\,g)\,\text{det}(U, V) &= \text{det}(\nabla_U(JG), V) - \text{det}(\nabla_V(JG), U) \\
&= \langle \nabla_V G, U \rangle - \langle \nabla_U G, V \rangle \\
&= d_V \langle G, U \rangle - d_U \langle G, V \rangle \\
&= d_V d_U g - d_U d_V g \\
&= 0.
\end{aligned}
$$

$\square$

The theorem below is a reformulation of Stokes Theorem in terms of vector fields instead of 1-forms. The integral

$$\int_{\partial M} g\,ds$$

of a function $g: \partial M \rightarrow \mathbb{R}$ is defined in the same way as for total geodesic curvature—as the sum of integrals over the boundary loops.

Theorem 12.7 (Divergence Theorem)
Let $Y \in \Gamma(TM)$ be a vector field and B the outward-pointing unit normal field on the boundary ∂M. Then

$$\int_M \text{div}\,Y\,\text{det} = \int_{\partial M} \langle Y, B \rangle\,ds.$$

Proof. Define a 1-form $\omega \in \Omega^1(M)$ by setting for $X \in T_p M$

$$\omega(X) = \langle JY(p), X \rangle.$$

Then, by Theorems 9.2, 9.4, 9.6 and Lemma 12.3,

$$d\omega(U, V) = d_U\omega(V) - d_V\omega(U)$$
$$= \langle J\nabla_U Y, V\rangle + \langle JY, \nabla_U V\rangle - \langle J\nabla_V Y, U\rangle - \langle JY, \nabla_V U\rangle$$
$$= \det(\nabla_U Y, V) - \det(\nabla_V Y, U)$$
$$= \mathrm{tr}(\nabla Y)\det(U, V).$$

Therefore $d\omega = \mathrm{div}\, Y \det$. Using again the notation of the proof of Theorem 10.6 and applying Stokes Theorem 7.15 we obtain

$$\int_M \mathrm{div}\, Y \det = \int_M d\omega$$
$$= \int_{\partial M} \omega$$
$$= \int_{\partial M} \langle JY, T\rangle ds$$
$$= \int_{\partial M} \langle Y, B\rangle ds.$$

$\square$

12.2   One-Parameter Families of Surfaces

Throughout this chapter $M \subset \mathbb{R}^2$ will be a compact domain with smooth boundary and $[t_0, t_1] \subset \mathbb{R}$ a closed interval.

Definition 12.8

Let $g_t : M \to \mathbb{R}^n$ a smooth map, defined for each $t \in [t_0, t_1]$. Then the one-parameter family of maps $[t_0, t_1] \ni t \mapsto g_t$ is called smooth if the map

$$M \times [t_0, t_1] \to \mathbb{R}^n, \quad (p, t) \mapsto g_t(p)$$

is smooth (as always, in the sense of Remark 1.2).

▶ **Remark 12.9** The variable t is also referred to as the **time**.

Given a smooth one-parameter family

$$t \mapsto (g_t : M \to \mathbb{R}^n), \quad t \in [t_0, t_1]$$

of maps and a vector field $X \in \Gamma(TM)$, also

$$t \mapsto d_X g_t$$

is a smooth one-parameter family of maps $d_X g_t : M \to \mathbb{R}^n$. The same holds for $t \mapsto \dot{g}_t$ where $\dot{g}_t : M \to \mathbb{R}^n$ is defined as

$$\dot{g}_t(p) := \frac{d}{d\tau}\Big|_{\tau=t} g_\tau(p).$$

The following fact will be used many times in upcoming chapters:

Theorem 12.10

For a smooth one-parameter family of maps $t \mapsto g_t$ from M to $\mathbb{R}^n$, the directional derivative in the direction of a vector field $X \in \Gamma(TM)$ commutes with the time derivative:

$$(d_X g_t)^\bullet = d_X \dot{g}_t.$$

Proof. In the special case where X is one of the coordinate vector fields U and V, this is just the fact that partial derivatives of the smooth map $(p, t) \mapsto g_t(p)$ commute. In the general case, we can write

$$X = a\, U + b\, V$$

where $a, b \in C^\infty(M)$ are independent of t. Then

$$\begin{aligned}
(d_X g_t)^\bullet &= (a\, d_U g_t + b\, d_V g_t)^\bullet \\
&= a\, d_U \dot{g}_t + b\, d_V \dot{g}_t \\
&= d_X \dot{g}_t.
\end{aligned}$$

$\square$

Definition 12.11

A smooth one-parameter family $t \mapsto g_t$ of maps from M to $\mathbb{R}^n$ is called a **variation of a smooth map** $g : M \to \mathbb{R}^n$ if

$$t_0 < 0 < t_1$$

and

$$g_0 = g.$$

In this context, we will also use the notation

$$\dot{g} := \dot{g}_0.$$

One should compare the arguments below with our reasoning in Sect. 2.4.

Definition 12.12

A **variation of a surface** $f: M \to \mathbb{R}^n$ is a smooth one-parameter family of surfaces

$$f_t: M \to \mathbb{R}^n, \quad t \in [-\epsilon, \epsilon]$$

such that

$$f_0 = f.$$

The map $\dot{f}: M \to \mathbb{R}^n$ defined as

$$\dot{f} := \dot{f}_0$$

is called the **variational vector field** of the variation $t \mapsto f_t$.

Definition 12.13

Let $M \subset \mathbb{R}^2$ be a compact domain with smooth boundary. Suppose we have a way to assign to each surface $f: M \to \mathbb{R}^n$ a real number $\mathcal{E}(f)$. Then $\mathcal{E}$ is called a **smooth functional** if for every smooth one-parameter family

$$t \mapsto f_t, \quad t \in [t_0, t_1]$$

of surfaces $f: M \to \mathbb{R}^n$ the function

$$[t_0, t_1] \to \mathbb{R}, \ t \mapsto \mathcal{E}(f_t)$$

is smooth.

In many circumstances, we want to consider only variations of $f: M \to \mathbb{R}^n$ that keep the surface fixed near the boundary ∂M:

Definition 12.14

Let $M \subset \mathbb{R}^2$ be a compact domain with smooth boundary and $f: M \to \mathbb{R}^n$ a surface. Then a variation

$$t \mapsto f_t, \quad t \in [-\epsilon, \epsilon]$$

of f is said to have **support in the interior** of M if there is a compact set $M_0 \subset \overset{\circ}{M}$ such that for all $p \in M$, $p \notin M_0$ we have

$$f_t(p) = f(p) \quad \text{for all} \quad t \in [-\epsilon, \epsilon].$$

Definition 12.15

Let $M \subset \mathbb{R}^2$ be a compact domain with smooth boundary and $\mathcal{E}$ be a smooth functional defined on the space of surfaces $f: M \to \mathbb{R}^n$. Then a surface $f: M \to \mathbb{R}^n$ is called a **critical point** of $\mathcal{E}$ if for all variations $t \mapsto f_t$ of f with support in the interior of M we have

$$\left. \frac{d}{dt} \right|_{t=0} \mathcal{E}(f_t) = 0.$$

Definition 12.15 spells out the notion of an equilibrium of a variational energy $\mathcal{E}$, to which we will refer to in later sections. Moreover, one should note that, as already explained in the beginning of Sect. 2.4, we will work with a definition of a critical point under constraints that is slightly stronger than the standard one.

Definition 12.16

Let $M \subset \mathbb{R}^2$ be a compact domain with smooth boundary, $f: M \to \mathbb{R}^n$ a surface and $\mathcal{E}, \tilde{\mathcal{E}}$ two smooth functionals on the space of all surfaces $\tilde{f}: M \to \mathbb{R}^n$. Then f is called a **critical point** of $\mathcal{E}$ **under the constraint** of fixed $\tilde{\mathcal{E}}$ if for all variations $t \mapsto f_t$ of f with support in the interior of M

$$\left. \frac{d}{dt} \right|_{t=0} \tilde{\mathcal{E}} = 0$$

implies

$$\left. \frac{d}{dt} \right|_{t=0} \mathcal{E} = 0.$$

Using the Linear Algebra Therorem 2.21 in the same way as we used it in Sect. 2.4, we obtain

Theorem 12.17
Let $M \subset \mathbb{R}^2$ be a compact domain with smooth boundary and $\mathcal{E}, \widetilde{\mathcal{E}}$ two smooth functionals on the space of all surfaces $f : M \to \mathbb{R}^n$. Suppose we have a way to associate to each surface $f : M \to \mathbb{R}^n$ smooth maps

$$G_f, \widetilde{G}_f : M \to \mathbb{R}^n$$

such that for all variations $t \mapsto f_t$ of f with support in the interior of M we have

$$\frac{d}{dt}\bigg|_{t=0} \mathcal{E} = \int_M \langle \dot{f}, G_f \rangle \det$$

$$\frac{d}{dt}\bigg|_{t=0} \widetilde{\mathcal{E}} = \int_M \langle \dot{f}, \widetilde{G}_f \rangle \det.$$

Then f is a critical point of $\mathcal{E}$ under the constraint of fixed $\widetilde{\mathcal{E}}$ if and only if there is a constant $\lambda \in \mathbb{R}$ such that

$$G_f = \lambda \widetilde{G}_f.$$

For reasons already explained in Sect. 2.4, we call λ a **Lagrange multiplier** for the constraint of fixed $\widetilde{\mathcal{E}}$.

12.3 Variation of Curvature

Given a smooth variation $t \mapsto f_t$ of a surface f, we are mainly interested in the time derivative at time zero of quantities like the area form $\det_t$ or the shape operator A_t associated with the surfaces f_t. In situations where it clear with which variation $t \mapsto f_t$ we are dealing, we will usually drop the index zero when we mean the time derivative at time zero. So, for example, we will write

$$\dot{A} = \dot{A}_0.$$

Theorem 12.18

Let $f \colon M \to \mathbb{R}^3$ be a surface with unit normal N, shape operator A and Levi-Civita connection ∇. Let $t \mapsto f_t$ be a variation of f whose variational vector field

$$\dot{f} = \phi N + df(Z)$$

is described in terms of a function $\phi \in C^\infty(M)$ and a vector field $Z \in \Gamma(TM)$. Denote by N_t and A_t the unit normals and the shape operators of the surfaces f_t. Define vector fields $G, W \in \Gamma(TM)$ as

$$G := \operatorname{grad} \phi$$

$$W := AZ - G.$$

Then

$$d\dot{f}(X) = -\langle W, X \rangle N + df(\phi AX + \nabla_X Z)$$

$$\dot{N} = df(W)$$

$$\dot{\det} = (2\phi H + \operatorname{div} Z)\det$$

$$\dot{A} = \nabla_Z A - \nabla G - \phi A^2 + A(\nabla Z) - (\nabla Z)A$$

$$\dot{H} = d_Z H - \frac{1}{2}\operatorname{div} G - \phi(2H^2 - K).$$

Proof. The proof of the first equation is straightforward:

$$d\dot{f}(X) = d\phi(X)N + \phi\, df(AX) - \langle AX, Z \rangle N + df(\nabla_X Z)$$
$$= \langle G - AZ, X \rangle N + df(\phi AX + \nabla_X Z)$$

Differentiating $\langle N, df(X) \rangle = 0$ with respect to time we obtain

$$\langle \dot{N}, df(X) \rangle = -\langle N, d\dot{f}(X) \rangle = \langle W, X \rangle = \langle df(W), df(X) \rangle.$$

This holds for all $X \in TM$ and this implies the second equation. For $X, Y \in T_p M$ we know that $\dot{N}$ (which is orthogonal to N), $df(X)$ and $df(Y)$ are linearly dependent. Using this and Theorem 12.3 we obtain

$$\dot{\det}(X, Y) = \det(N, df(X), df(Y))^\bullet$$

$$= \det(\dot{N}, df(X), df(Y)) + \det(N, df(\phi AX + \nabla_X Z), df(Y))$$

$$+ \det(N, df(X), df(\phi AY + \nabla_Y Z))$$

$$= \det(\phi AX + \nabla_X Z, Y) + \det(X, \phi AY + \nabla_Y Z)$$

$$= \mathrm{tr}(\phi A + \nabla Z) \det(X, Y)$$

$$= (2\phi H + \mathrm{div}\, Z) \det(X, Y).$$

This proves the third equation. For the fourth equation, consider the directional derivative of the second equation in the direction of X and make use of the first:

$$-\langle AX, W \rangle N + df(\nabla_X W) = d\dot{N}(X)$$

$$= (dN(X))^\bullet$$

$$= (df(AX))^\bullet$$

$$= d\dot{f}(AX) + df(\dot{A}X)$$

$$= -\langle W, AX \rangle N + df(\phi A^2 X + \nabla_{AX} Z) + df(\dot{A}X)$$

The normal part of this equation is satisfied automatically. The tangential part, together with the Codazzi equation (Theorem 9.7) gives us

$$\dot{A}X = \nabla_X W - \phi A^2 X - \nabla_{AX} Z$$

$$= \nabla_X(AZ) - \nabla_X G - \phi A^2 X - \nabla_{AX} Z$$

$$= (\nabla_X A)Z + A\nabla_X Z - \nabla_X G - \phi A^2 X - \nabla_{AX} Z$$

$$= (\nabla_Z A)X + A(\nabla Z)(X) - (\nabla G)(X) - \phi A^2 X - (\nabla Z)(AX).$$

This proves the fourth equation. For the fifth we take the trace of the fourth and multiply by $\frac{1}{2}$. The last two terms in the fourth equation do not contribute because we see here the commutator of two endomorphisms A and ∇Z, which always has zero trace. Regarding the first term, one can verify (for example by taking the directional derivative of the equation in Theorem 12.3 in the direction of Z) that indeed for any endomorphism field $\tilde{A}$

$$\mathrm{tr}(\nabla_Z \tilde{A}) = d_Z(\mathrm{tr}\,\tilde{A}).$$

Finally, by diagonalizing A one can easily check the equality

$$\frac{1}{2}\operatorname{tr} A^2 = 2H^2 - K.$$

$\square$

12.4   Variation of Area

Variations of surfaces (as defined in Definition 12.12) are needed in order to define and determine those surfaces that represent equilibria of geometrically interesting variational functionals.

Examples of smooth functionals of surfaces are the Willmore functional (to be introduced in Sect. 13.1) and the cone volume that will be defined in Sect. 12.5. In this chapter we will focus on the area functional

$$\mathcal{A}(f) = \int_M \det{}_f.$$

Theorem 12.19 (First Variation Formula of Area)
As in Theorem 12.18, suppose the variational vector field of a variation $t \mapsto f_t$ of a surface $f: M \to \mathbb{R}^3$ is written as

$$\dot{f} = \phi N + df(Z)$$

with $\phi \in C^\infty(M)$ and $Z \in \Gamma(TM)$. Then

$$\left.\frac{d}{dt}\right|_{t=0} \mathcal{A}(f_t) = 2 \int_M \phi H \det + \int_{\partial M} \langle Z, B \rangle \, ds$$

where B is the outward pointing unit normal on ∂M.

Proof. By Theorem 12.18 and the Divergence Theorem 12.7,

$$\left.\frac{d}{dt}\right|_{t=0} \mathcal{A}(f_t) = \int_M \dot{\det}$$

$$= 2 \int_M \phi H \, \det + \int_M \operatorname{div} Z \, \det$$

$$= 2 \int_M \phi H \, \det + \int_{\partial M} \langle Z, B \rangle \, ds.$$

$\square$

Fig. 12.2 The **Schwarz-P**
minimal surface

Definition 12.20

A surface $f\colon M \to \mathbb{R}^3$ is called a **minimal surface** if it is a critical point of the
area functional $\mathcal{A}$.

Figure 12.2 shows a minimal surface whose six boundary curves are all mapped onto
prescribed circles. In fact, it is here a solution of the so-called **Plateau problem**,
which means that it minimizes area among all surfaces whose boundary is mapped
onto a prescribed set of curves.

▶ **Remark 12.21** The Plateau problem was first solved by Jesse Douglas [12] and
Tibor Rado [32] independently.

Theorem 12.22
*A surface $f\colon M \to \mathbb{R}^3$ is a minimal surface if and only if is mean curvature
H vanishes.*

Proof. If $H = 0$ and $t \mapsto f_t$ is a variation of f with support in the interior of M,
then Z vanishes near the boundary of M and by Theorem 12.19 the variation of area
is zero. Conversely, suppose that f is a minimal surface but there is a point $p \in M$
for which $H(p) \neq 0$. Then there is such a p also in the interior of M, so we assume
$p \in \overset{\circ}{M}$. Let us treat the case $H(p) > 0$, the case $H(p) < 0$ being similar. Then
we can construct a bump function $g \in C^\infty(M)$ such that g vanishes outside of a
compact set contained in the interior of M and

$$g(p) = 1$$

$$H(q) \leq 0 \implies g(q) = 0.$$

Fig. 12.3 An **Enneper surface** is a minimal surface

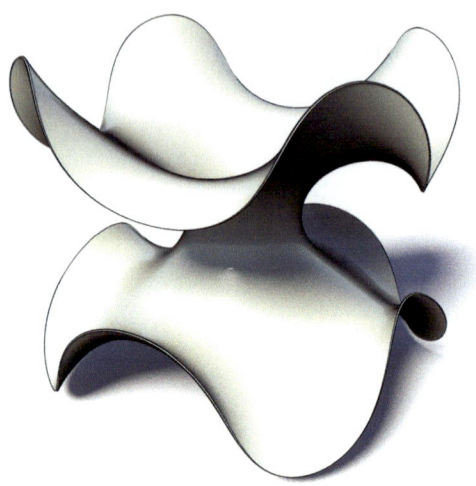

Then, for small enough $\epsilon > 0$,

$$t \mapsto f_t, \quad t \in [-\epsilon, \epsilon]$$

$$f_t = f + t \cdot g \cdot N$$

(N being the unit normal of f) will be a smooth variation of f with support in the interior of M and

$$\frac{d}{dt}\bigg|_{t=0} \mathcal{A}(f_t) = \int_M gH > 0,$$

which contradicts our assumption that f is minimal.   □

As the reader may verify, the Enneper surfaces defined in Sect. 6.5 have mean curvature $H = 0$, so by Theorem 12.22 they are minimal surfaces. Figure 12.3 shows one of these Enneper surfaces.

12.5   Variation of Volume

Definition 12.23

Let $M \subset \mathbb{R}^2$ be a compact domain with smooth boundary and $f \colon M \to \mathbb{R}^3$ a surface. Then the **cone volume** of f is defined as

$$V(f) = \frac{1}{3} \int_M \det(f, f_u, f_v).$$

$\mathcal{V}(f)$ can be interpreted as the volume covered by the map

$$F : [0, 1] \times M \to \mathbb{R}^3, \quad F(s, p) = s \cdot f(p).$$

Here the "volume covered" should not be understood as the volume of the image $F([0, 1] \times M)$, but rather in the spirit of Theorem 8.17. At first sight, the cone volume does not look like an honorable geometric functional. For example, the version $\tilde{f} = f + \mathbf{a}$ of f that has been translated by a vector $\mathbf{a} \in \mathbb{R}^3$ in general does not have the same cone volume as f. On the other hand, for closed surfaces the cone volume is invariant under translations:

Theorem 12.24

If (f, ρ) is an oriented closed surface (Definition 11.1) and $\mathbf{a} \in \mathbb{R}^3$, then

$$\mathcal{V}(f + \mathbf{a}) = \mathcal{V}(f).$$

Proof. Define a 1-form $\omega \in \Omega^1(M)$ by

$$\omega(X) = \frac{1}{6} \det(\mathbf{a}, f, df(X)).$$

Then

$$d\omega(U, V) = \frac{1}{6}(\det(\mathbf{a}, f, f_v)_u - \det(\mathbf{a}, f, f_u)_v = \frac{1}{3} \det(\mathbf{a}, f_u, f_v)$$

and therefore, by Stokes Theorem 7.15,

$$\mathcal{V}(f + \mathbf{a}) - \mathcal{V}(f) = \int_M d\omega = \int_{\partial M} \omega = 0.$$

The last equality follows from the fact that (f, ρ) is oriented, and therefore the integrals of ω over the various boundary curves of M cancel in pairs.   □

Fig. 12.4 The cone volume
of f *(left)* and of a variation
f_t of f *(right)*

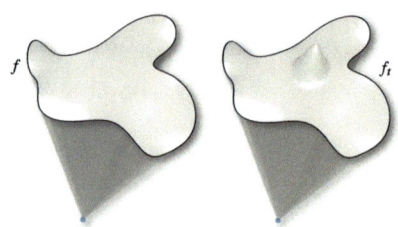

Moreover, by almost the same reasoning as in the above proof one can show:

Theorem 12.25
Let $M \subset \mathbb{R}^2$ be a compact domain with smooth boundary,

$$t \mapsto f_t, \quad t \in [-\epsilon, \epsilon]$$

a variation with support in the interior of M of a surface $f : M \to \mathbb{R}^3$ and $\mathbf{a} \in \mathbb{R}^3$. Then

$$\frac{d}{dt}\bigg|_{t=0} \mathcal{V}(f_t + \mathbf{a}) = \frac{d}{dt}\bigg|_{t=0} \mathcal{V}(f_t).$$

Theorem 12.25 implies that for the purposes of variational calculus the cone volume $\mathcal{V}$ behaves in the same way as a translationally invariant functional (see Fig. 12.4).

We can view df as an $\mathbb{R}^3$-valued 1-form on M. Given smooth maps $f, \dot{f} : M \to \mathbb{R}^3$ we then obtain a scalar valued 1-form

$$\omega = \frac{1}{3} \det(f, \dot{f}, df) \in \Omega^1(M).$$

Theorem 12.26 (First Variation of Cone Volume)
Let $f : M \to \mathbb{R}^3$ be a surface. Then for every variation $t \mapsto f_t$ of f we have

$$\frac{d}{dt}\bigg|_{t=0} \mathcal{V}(f_t) = \int_M \langle \dot{f}, N \rangle \det + \int_{\partial M} \frac{1}{3} \det(f, \dot{f}, df).$$

Proof. We have

$$
\frac{d}{dt}\Big|_{t=0} \mathcal{V}(f_t) = \frac{1}{3} \int_M \left(\det(\dot{f}, f_u, f_v) + \det(f, \dot{f}_u, f_v) + \det(f, f_u, \dot{f}_v) \right)
$$

$$
= \frac{1}{3} \int_M \left(\det(\dot{f}, f_u, f_v) + \det(f, \dot{f}, f_v)_u - \det(f_u, \dot{f}, f_v) \right.
$$

$$
- \det(f, \dot{f}, f_{vu}) + \det(f, f_u, \dot{f})_v - \det(f_v, f_u, \dot{f})
$$

$$
\left. - \det(f, f_{uv}, \dot{f}) \right)
$$

$$
= \int_M \langle \dot{f}, \det(U, V) N \rangle + \int_M d\omega
$$

$$
= \int_M \langle \dot{f}, N \rangle \det + \int_{\partial M} \omega.
$$

$\square$

It is easy to see that, on its own, the cone volume functional does not have any critical points. However, we can use it in the context of variational problems under a volume constraint. Here is our first application of Theorem 12.17:

> **Theorem 12.27**
> *Let $M \subset \mathbb{R}^2$ be a compact domain with smooth boundary. Then a surface $f: M \rightarrow \mathbb{R}^3$ is a critical point of the area $\mathcal{A}$ under the constraint of fixed cone volume $\mathcal{V}$ if and only if the mean curvature H of f is constant.*

Proof. By Theorems 12.19, 12.26, and 12.17, f is a critical point of area under fixed cone volume if and only if there is a constant $\lambda \in \mathbb{R}$ such that

$$
HN = \lambda N.
$$

$\square$

The surface in Fig. 12.5 minimizes area among all surfaces that are bounded by the same six circles as the first surface shown in Sect. 12.4 and have a certain prescribed volume:

Fig. 12.5 A surface with the same boundary as the surface in Fig. 12.2. It is a critical point of the area functional under the constraint of having a prescribed cone volume

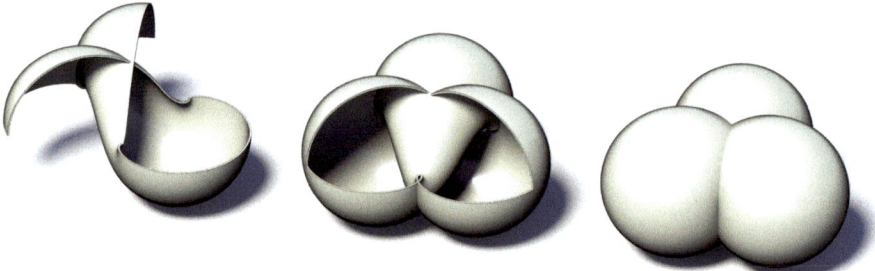

Fig. 12.6 A Wente torus—a closed surface of genus $g = 1$ with constant mean curvature $H \neq 0$

▶ **Remark 12.28** In 1984 Henry Wente found a counterexample to a conjecture by Heinz Hopf which stated that every closed surface in $\mathbb{R}^3$ with constant mean curvature is round sphere [43]. In Fig. 12.6 it is shown how the **Wente torus** can be build from a fundamental piece.

Nevertheless, the conjecture is true if one demands that the surface is embedded in $\mathbb{R}^3$, or has genus $g = 0$. These results are due to Alexandrov [1] and Hopf [17].

Willmore Surfaces

13

The analog for a surface $f : M \to \mathbb{R}^3$ of the bending energy $\int_a^b \kappa^2\, ds$ is the *Willmore functional* $W(f) = \int_M H^2 \det$. There are several versions of the Willmore functional, all of which are equivalent for the purposes of Variational Calculus. One of these versions is unchanged if we transform the surface by inversion in a sphere. The analogs of elastic curves are called *Willmore Surfaces*.

13.1 The Willmore Functional

In the context of curves $\gamma : [a, b] \to \mathbb{R}^2$ we studied in detail the total squared curvature $\int \kappa^2 ds$ (notation from the end of Sect. 7.2). What is the analog of this energy in the context of surfaces?

One might say that $\kappa = 0$ characterizes straight lines, which minimize length among all curves with the same end points. So $\int_{[a,b]} \kappa^2\, ds$ measures the deviation from being length-minimizing. The analog of length-minimizing curves are area-minimizing surfaces, i.e. minimal surfaces, surfaces with mean curvature $H = 0$. So a natural analog of $\int_{[a,b]} \kappa^2\, ds$ can be defined as follows:

Definition 13.1

If $f : M \to \mathbb{R}^3$ is a surface, then

$$W(f) := \int_M H^2 \det$$

is called the **Willmore functional** of f.

© The Author(s) 2024
U. Pinkall, O. Gross, *Differential Geometry*, Compact Textbooks in Mathematics,
https://doi.org/10.1007/978-3-031-39838-4_13

Surfaces f that are critical points of the Willmore functional are characterized by the property that they are "as minimal as possible", given that they are held fixed near the boundary of M.

Alternatively, one might say that $\kappa = 0$ only happens for straight line segments, so for surfaces we want to measure the deviation of being planar. Parametrizations of pieces of the plane are characterized by the fact that both principal curvatures vanish, so we want to measure the deviation of both κ_1 and κ_2 (not just their average H) from being zero. This reasoning leads to a different analog for $\int_{[a,b]} \kappa^2 \, ds$:

Definition 13.2

If $f \colon M \to \mathbb{R}^3$ is a surface, then

$$E(f) := \frac{1}{4} \int_M (\kappa_1^2 + \kappa_2^2) \det = \int_M \left(H^2 - \frac{K}{2} \right) \det$$

is called the **bending energy** of f.

Surfaces f that are critical points of the bending energy are characterized by the property that they are "as planar as possible", given that they are held fixed near the boundary.

Finally, one might formulate a different wish and ask for surfaces that are "as round as possible" which means they are "as spherical as possible". In view of the Umbillic Point Theorem 8.12 this motivates the following definition:

Definition 13.3

If $f \colon M \to \mathbb{R}^3$ is a surface, then

$$\widetilde{W}(f) := \frac{1}{4} \int_M (\kappa_1 - \kappa_2)^2 \det = \int_M (H^2 - K) \det$$

is called the **conformally invariant Willmore functional** of f.

Note that the integrands in all three of the above energies differ only by a term proportional to $K \det$, so the Gauss-Bonnet Theorem 10.6 tells us that for the purposes of Variational Calculus, (cf. Definition A.4) all three energies are equivalent to a large extent:

Theorem 13.4

Let $f, \tilde{f} \colon M \to \mathbb{R}^3$ be two surfaces such that $\tilde{f}(p) = f(p)$ for p outside of some compact set contained in the interior of M. Then

$$W(\tilde{f}) - W(f) = E(\tilde{f}) - E(f) = \widetilde{W}(\tilde{f}) - \widetilde{W}(f).$$

For surfaces that close up we see that the difference between the three functionals only depends on the genus:

Theorem 13.5
Let $f : M \to \mathbb{R}^3$ be a surface that closes up with genus g. Then

$$E(f) = W(f) + 2\pi(g - 1)$$

$$\widetilde{W}(f) = W(f) + 4\pi(g - 1).$$

Theorem 13.6
The estimates below are sharp, i.e. in each case there is a surface that closes up with the prescribed genus and which realizes the lower bound:

(i) If M is connected and a surface $f : M \to \mathbb{R}^3$ closes up with genus 0, then

$$W(f) \geq 4\pi.$$

(ii) If M is connected and a surface $f : M \to \mathbb{R}^3$ closes up with genus $\frac{1}{2}$, then

$$W(f) \geq 12\pi.$$

(iii) If M is connected and a surface $f : M \to \mathbb{R}^3$ closes up with genus 1, then

$$W(f) \geq 2\pi^2.$$

We will not prove this theorem. Part (i) of Theorem 13.6 was proved by Tom Willmore in [46] in 1965. The minimum is attained for a round sphere. Part (ii) was proved by Rob Kusner in [21] where he also proved that the Boy surface shown on the right of Fig. 13.1 realizes the minimum 12π. The two surfaces on the right of Fig. 13.2 are **Lawson surfaces** which were found by Blaine Lawson [23] and are possible candidates for minimizing the Willmore functional among all surfaces with genus $g = 2$ and $g = 3$ respectively [18].

In the paper already mentioned above, Willmore also formulated (iii) as a conjecture and demonstrated that the value $2\pi^2$ is realized by the torus obtained by rotating a circle of radius one around an axis in such a way that its center has distance $\sqrt{2}$ from the axis (Fig. 13.2, left). This **Willmore conjecture** remained

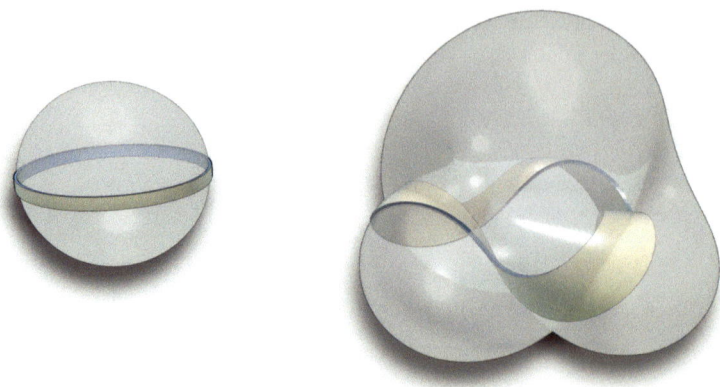

Fig. 13.1 A Boy surface with the minimal possible Willmore functional 12π

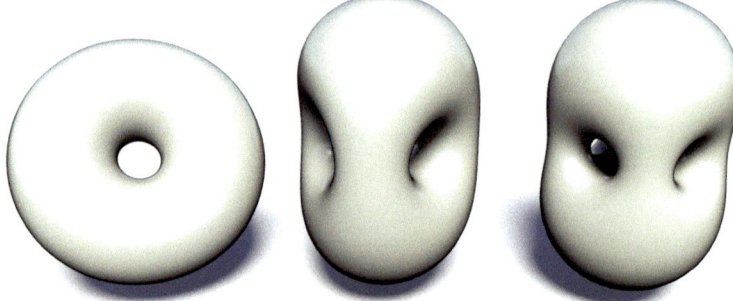

Fig. 13.2 The Torus on the left has Willmore functional $2\pi^2$, which is optimal among surfaces with genus $g = 1$. The two surfaces on the right are possible candidates for minimizing the Willmore functional among all surfaces with genus $g = 2$ and $g = 3$ respectively

a famous open problem in Differential Geometry for a long time, until in 2012 Fernando Marques and André Neves proved the conjecture [26].

▶ **Remark 13.7** The question of critical points of the Willmore functional acquired greater importance starting from the 1960s, initiated by T. Willmore and his paper [46]. It was later found that parts of the theory were already known to Wilhelm Blaschke [6] and his student Gerhard Thomsen in the 1920s [38]. For an historic overview of contributions which were made to the problem see the last chapter of [47], or [27] for a more recent survey.

13.2 Variation of the Willmore Functional

According to the discussion in Sect. 13.1, the Willmore functional $\mathcal{W}$ has alternative versions which measure how "non-flat" or how "not round" a surface is. It was also explained that for the purposes of Variational Calculus all these different versions of the Willmore functional are equivalent. Being a critical point of the Willmore functional (which version we take does not matter) means that the surface (at least locally) is "optimally round". It also means that the total amount of curvature of the surface cannot be decreased by modifying f only in a small neighborhood of a given point, while leaving the rest of the surface unchanged.

Definition 13.8

A surface $f : M \to \mathbb{R}^3$ is called a **Willmore surface** if it is a critical point of the Willmore functional $\mathcal{W}$.

Let us first compute for $\mathcal{W}$ the rate of change under a general variation, not necessarily with support in the interior:

Theorem 13.9 (First Variation Formula for the Willmore Functional)
Let $f : M \to \mathbb{R}^3$ be a surface with unit normal N and with binormal field B along the boundary ∂M. Let $t \mapsto f_t$ be a variation of f with variational vector field

$$\dot{f} = \phi N + df(Z)$$

where $\phi \in C^\infty(M)$ and $Z \in \Gamma(TM)$. Then

$$\frac{d}{dt}\Big|_{t=0} \mathcal{W}(f_t) = -\int_M \phi\left(\Delta H + 2H(H^2 - K)\right) \det$$

$$+ \int_{\partial M} \left\langle B, H^2 Z - H \operatorname{grad} \phi + \phi \operatorname{grad} H \right\rangle ds.$$

Proof. Using Theorem 12.18 as well as the notation $G := \operatorname{grad} \phi$ borrowed from there we obtain

$$\left(H^2 \det\right)^{\bullet} = \left(2H\left(d_Z H - \frac{1}{2}\operatorname{div} G - \phi(2H^2 - K)\right) + H^2(2H\phi + \operatorname{div} Z)\right) \det$$

$$= \left(2H\langle \operatorname{grad} H, Z\rangle - H \operatorname{div} G - 2\phi H(H^2 - K) + H^2\operatorname{div} Z\right) \det$$

$$= \left(\operatorname{div}(H^2 Z) - H \operatorname{div} G - 2\phi H(H^2 - K)\right) \det$$

$$= \Big(\operatorname{div}(H^2 Z - HG) + \langle \operatorname{grad} H, \operatorname{grad} \phi \rangle - 2\phi H (H^2 - K) \Big) \det$$

$$= \Big(\operatorname{div}(H^2 Z - HG) + \operatorname{div}(\phi \operatorname{grad} H) - \phi \Delta H - 2\phi H (H^2 - K) \Big) \det$$

$$= \Big(\operatorname{div} \Big(H^2 Z - HG + \phi \operatorname{grad} H \Big) - \phi \Big(\Delta H + 2H (H^2 - K) \Big) \Big) \det .$$

Together with the Divergence Theorem 12.7, this proves our claim.   □

As an immediate consequence, we obtain (cf. [38])

Theorem 13.10
A surface $f : M \to \mathbb{R}^3$ is a Willmore surface if and only if

$$\Delta H + 2H (H^2 - K) = 0.$$

Round spheres are Willmore, because for them all points are umbilic points (so $H^2 - K = 0$) and H is constant (so $\Delta H = 0$). Moreover, all surfaces with $H = 0$ (minimal surfaces) are Willmore. Here is another example:

Example 13.11
Take a unit speed curve $\gamma : [0, L] \to \mathbb{R}^2 \subset \mathbb{R}^3$, where $\mathbb{R}^2$ is realized as those points in $\mathbb{R}^3$ where the last coordinate is zero. Now for a compact domain with smooth boundary $M \subset [0, L] \times \mathbb{R}$ define the **cylinder** $f : M \to \mathbb{R}^3$ over γ by

$$f(u, v) = \gamma(u) + v\, e_3.$$

It is easy to check that the Levi-Civita connection of f is given by $\nabla U = \nabla V = 0$, the Gaussian curvature K of f vanishes and the mean curvature H of f satisfies

$$H(u, v) = \frac{\kappa(u)}{2}$$

$$(\operatorname{grad} H)(u, v) = \frac{\kappa'(u)}{2} U(u, v)$$

$$(\Delta H)(u, v) = \frac{\kappa''(u)}{2} .$$

This means that the cylinder f over γ is Willmore if and only if γ is freely elastic, i.e.

$$\kappa'' + \frac{\kappa^3}{2} = 0.$$

The cylinder over a freely elastic curve is seen in Fig. 13.3.

Fig. 13.3 The **cylinder over a free elastic plane curve** is a Willmore surface

Fig. 13.4 Another Willmore surface

There are many other ways to construct Willmore surfaces, most of which are beyond the scope of this book. The surface in Fig. 13.4 is from the 2019 paper [7].

13.3   Willmore Functional Under Inversions

For a surface $f \colon M \to \mathbb{R}^3$, the Willmore functional

$$\mathcal{W}(f) = \int_M H^2 \det$$

is clearly unchanged if we postcompose f by an isometry $g \colon \mathbb{R}^3 \to \mathbb{R}^3$. It is also invariant under scaling: For $\lambda \neq 0$ the surface $\tilde{f} = \lambda f$ has the same Willmore functional. This is because, under such a scaling, det acquires a factor of λ^2 while H gets a factor of $\frac{1}{\lambda}$. As its name indicates, if we consider the Möbius-invariant

Fig. 13.5 The images under an inversion of the surfaces in Figs. 13.3 and 6.8 respectively are also Willmore surfaces

Willmore functional

$$\widetilde{\mathcal{W}}(f) = \int_M (H^2 - K) \det$$

a similar statement is true for a more general class of transformations, that can be written as compositions of isometries, scalings and inversions in spheres, the so-called **Möbius transformations** (Fig. 13.5).

Let $f : M \to \mathbb{R}^3$ be a surface such that the origin $\mathbf{o}$ of $\mathbb{R}^3$ is not in the image of f. Then we can postcompose f with the so-called inversion in the unit sphere

$$g : \mathbb{R}^3 \setminus \{\mathbf{o}\} \to \mathbb{R}^3, \; g(\mathbf{p}) = \frac{\mathbf{p}}{\langle \mathbf{p}, \mathbf{p} \rangle}$$

and obtain a new surface

$$\tilde{f} : M \to \mathbb{R}^3, \; \tilde{f} = \frac{f}{\langle f, f \rangle}.$$

Computing the derivative of $\tilde{f}$ is straightforward and yields

$$d\tilde{f} = \frac{df}{\langle f, f \rangle} - 2 \frac{\langle df, f \rangle f}{\langle f, f \rangle^2}$$

$$= \frac{1}{\langle f, f \rangle} R \, df$$

where for each $p \in M$ the orthogonal (3×3)-matrix $R(p) \in O(3)$ acts on $\mathbf{v} \in \mathbb{R}^3$ as

$$R(p)\mathbf{v} = \mathbf{v} - 2\frac{\langle f(p), \mathbf{v}\rangle}{\langle f(p), f(p)\rangle}f(p).$$

For each $p \in M$ the matrix $R(p)$ is a reflection and hence orientation-reversing. The sign of the unit normal depends on orientation, which is why the unit normal field of $\tilde{f}$ is given by

$$\tilde{N} = -RN = \frac{2\langle N, f\rangle}{\langle f, f\rangle}f - N.$$

Theorem 13.12

In the situation above, the induced metric $\langle , \rangle^{\sim}$, the area form $\widetilde{\det}$ and the shape operator $\tilde{A}$ of $\tilde{f}$ are given by

$$\langle , \rangle^{\sim} = \frac{1}{\langle f, f\rangle^2}\langle , \rangle$$

$$\widetilde{\det} = \frac{1}{\langle f, f\rangle^2}\det$$

$$\tilde{A} = -\langle f, f\rangle A + 2\langle N, f\rangle I.$$

Proof. The first two formulas follow directly from our calculations above. The third follows from

$$d\tilde{N} = \left(\frac{2\langle dN, f\rangle}{\langle f, f\rangle} - \frac{4\langle N, f\rangle\langle df, f\rangle}{\langle f, f\rangle^2}\right)f + \frac{2\langle N, f\rangle}{\langle f, f\rangle}df - dN$$

$$= 2\langle N, f\rangle d\tilde{f} - \langle f, f\rangle d\tilde{f} \circ A.$$

$\square$

Theorem 13.13

If $\tilde{f}$ arises from f by inversion in the unit sphere, then

$$\widetilde{\mathcal{W}}(\tilde{f}) = \widetilde{\mathcal{W}}(f).$$

Proof. By Theorem 13.12, the principal curvatures $\tilde{\kappa}_1, \tilde{\kappa}_2$ of $\tilde{f}$ satisfy

$$\tilde{\kappa}_2 - \tilde{\kappa}_1 = -\langle f, f \rangle (\kappa_2 - \kappa_1).$$

As a consequence,

$$(\tilde{H}^2 - \tilde{K})\widetilde{\det} = \frac{1}{4}(\tilde{\kappa}_2 - \tilde{\kappa}_1)^2 \, \widetilde{\det} = \frac{1}{4}(\kappa_2 - \kappa_1)^2 \det = (H^2 - K) \det .$$

$\square$

Theorem 13.14
If f is a Willmore surface, then so is its image $\tilde{f}$ under inversion in the unit sphere.

Proof. By Theorem 13.4, $\widetilde{\mathcal{W}}$ has the same critical points as $\mathcal{W}$ and by Theorem 13.13 inversion in the unit sphere maps critical points of $\widetilde{\mathcal{W}}$ to critical points of $\widetilde{\mathcal{W}}$.

$\square$

The surface on the right of Fig. 13.6 shows the image under an inversion of a minimal surface already known to Euler (shown on the left), the so-called **catenoid** $f : M \to \mathbb{R}^3$ given by

$$f(u, v) = \begin{pmatrix} \frac{1+u^2+v^2}{u^2+v^2} u \\ \frac{1+u^2+v^2}{u^2+v^2} v \\ \log(u^2 + v^2) \end{pmatrix}.$$

So, even more Willmore surfaces can be obtained by inverting surfaces which we have already encountered (see Fig. 13.5).

Fig. 13.6 A **catenoid** *(left)* and its image under a sphere inversion *(right)*

Some Technicalities

A

A.1   Smooth Maps

The standard definition of a differentiable map $f \colon U \to \mathbb{R}^n$ requires U to be an open subset of $\mathbb{R}^k$. Such an f is called smooth if all higher order partial derivatives

$$\frac{\partial^m f_i}{\partial x_{j_1} \dots \partial x_{j_m}}$$

of all its component functions exist.

On the other hand, we want to define curves in $\mathbb{R}^n$ as certain smooth maps $\gamma \colon [a, b] \to \mathbb{R}^n$ defined on a closed interval $[a, b] \subset \mathbb{R}$. Similarly, we want to use a certain kind of compact subsets $M \subset \mathbb{R}^2$ as the domain of definition for surfaces $f \colon M \to \mathbb{R}^n$. We therefore have to work with more general domains:

Definition A.1

Let $U \subset \mathbb{R}^k$ be an open set and $M := \overline{U}$ its closure. Then a function $f \colon M \to \mathbb{R}^n$ is called **smooth** if there is an open set $\tilde{U} \subset \mathbb{R}^n$ with $M \subset \tilde{U}$ and a smooth function $\tilde{f} \colon \tilde{U} \to \mathbb{R}^n$ such that

$$\tilde{f}|_M = f.$$

For $x \in M$ we define

$$\frac{\partial^m f_i}{\partial x_{j_1} \dots \partial x_{j_m}} := \frac{\partial^m \tilde{f}_i}{\partial x_{j_1} \dots \partial x_{j_m}}$$

© The Author(s) 2024
U. Pinkall, O. Gross, *Differential Geometry*, Compact Textbooks in Mathematics,
https://doi.org/10.1007/978-3-031-39838-4

In order for this definition to make sense, we have to verify that the higher partial derivatives of f are well-defined:

Theorem A.2

The higher partial derivatives of f defined in Definition A.1 are independent of the choice $\tilde{U}$ and the extension $\tilde{f}$ of f to $\tilde{U}$.

Proof. Because every point $x \in M$ is a limit point of points $y \in U$ and the partial derivatives of $\tilde{f}$ are continuous, we have

$$\frac{\partial^m \tilde{f}_i}{\partial x_{j_1} \dots \partial x_{j_m}}(x) = \lim_{\substack{y \to x \\ y \in U}} \frac{\partial^m \tilde{f}}{\partial x_{j_1} \dots \partial x_{j_m}}(y)$$

$$= \lim_{\substack{y \to x \\ y \in U}} \frac{\partial^m f}{\partial x_{j_1} \dots \partial x_{j_m}}(y).$$

$\square$

When we discuss reparametrizations of curves or surfaces, we make use of the following notion:

Definition A.3

Let $M, \tilde{M} \subset \mathbb{R}^k$ be two subsets which are closures of open subsets $U, \tilde{U} \subset \mathbb{R}^n$ respectively. Then a smooth map $f : M \to \tilde{M}$ is called a **diffeomorphism** if it is bijective and its inverse $f^{-1} : \tilde{M} \to M$ is also smooth.

A.2   Function Toolbox

On several occasions in this book the need arises to construct a so-called bump function, i.e. a non-negative smooth function on $\mathbb{R}^k$ that vanishes outside of a small neighborhood of a given point, but not at this point.

Definition A.4

Let $f : A \to \mathbb{R}$ be a function defined on some subset $A \subset \mathbb{R}^k$ of $\mathbb{R}^k$. Then the **support** of f is defined as

$$\text{supp } f := \{x \in \mathbb{R}^n \mid \forall \epsilon > 0 \; \exists \, y \in A \text{ with } |y - x| < \epsilon \text{ and } f(y) \neq 0\}.$$

The following theorem is easy to prove.

> **Theorem A.5**
> *If $f : U \to \mathbb{R}$ is a smooth function on an open set $U \subset \mathbb{R}^k$ and supp $f \subset U$, then f can be extended to a smooth function $\tilde{f} : \mathbb{R}^k \to \mathbb{R}$ by setting*
>
> $$\tilde{f}(x) := \begin{cases} f(x) & \text{for } x \in U \\ 0 & \text{for } x \notin U. \end{cases}$$

The basic ingredient for constructing functions with support in a given open set $U \subset \mathbb{R}^k$ is the function $f : \mathbb{R} \to \mathbb{R}$ given by

$$f(x) = \begin{cases} 0 & \text{for } x \leq 0 \\ e^{-\frac{1}{x}} & \text{for } x > 0. \end{cases}$$

Clearly, f is smooth at all points $x \neq 0$. It is not hard to check that it is smooth also at $x = 0$. Figure A.1 shows the graph of f.

The second function in our toolbox is the so-called **bump function** $g : \mathbb{R} \to \mathbb{R}$ given by

$$g(x) = f(1 - x^2) = \begin{cases} 0 & \text{for } |x| \geq 1 \\ e^{-\frac{1}{1-x^2}} & \text{for } |x| < 1. \end{cases}$$

As a composition of smooth functions, g is also smooth (see Fig. A.2).

Other versions of g like

$$\tilde{g}(x) = g(\epsilon(x - x_0))$$

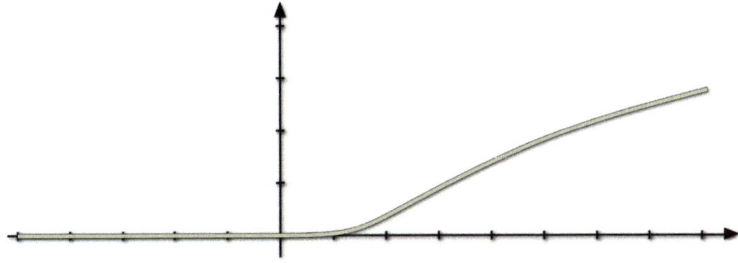

Fig. A.1 A smooth fuction $f : \mathbb{R} \to \mathbb{R}$ with $f(x) = 0$ for $x < 0$

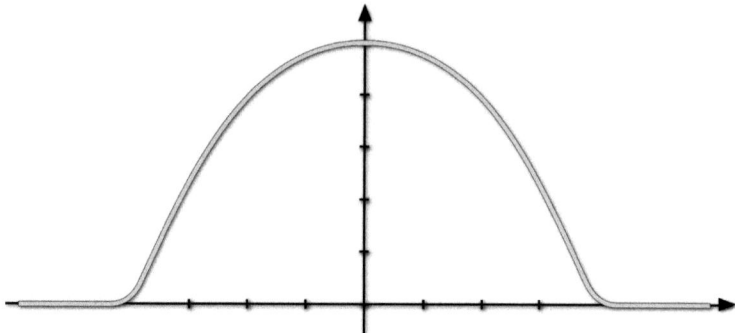

Fig. A.2 The bump function $g \colon \mathbb{R} \to \mathbb{R}$

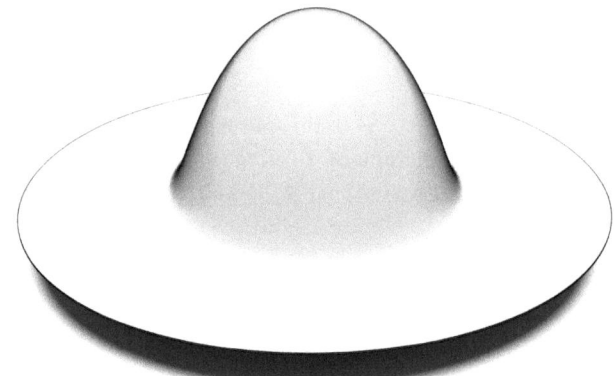

Fig. A.3 A bump function $\hat{g}$ on $\mathbb{R}^2$

can be adapted to be non-zero only within an arbitrarily prescribed interval. Another tool in our toolbox is the function $h \colon \mathbb{R} \to \mathbb{R}$ given by

$$h(x) = \int_{-1}^{x} g$$

and variants of it that are shifted and scaled in a similar way as the function $\tilde{g}$ above.

Finally, we need bump functions of several variables (see Fig. A.3), like the function $\hat{g} \colon \mathbb{R}^k \to \mathbb{R}$ given by

$$\hat{g}(x) = f(1 - \|x\|^2).$$

Timeline

B

Table B.1 is not meant as a comprehensive overview of the history of the whole field that deals with the Differential Geometry of curves and surfaces. Only those milestones are listed that are explicitly mentioned in the preceding chapters.

Table B.1 The milestones of differential geometry of curves and surfaces which are covered in this book

Year	Milestone	Section
1673	Newton defines the curvature of curves in $\mathbb{R}^2$	3.1
1691	Jacob Bernoulli defines elastic curves in $\mathbb{R}^2$	2.4
1744	Euler classifies elastic curves in $\mathbb{R}^2$	2.5
1744	Euler shows that the catenoid minimizes area	13.3
1760	Euler defines the principal curvatures of a surface	8.2
1827	Gauss proves the Theorema Egregium	9.3
1844	Binet derives the equation of elastic curves in $\mathbb{R}^3$	5.4
1845	Möbius investigates the topology of closed surfaces	11.1
1848	Bonnet proves the Gauss-Bonnet theorem	10.2
1859	Kirchhoff proves that the tangent of an elastic curve follows the motion of the axis of a spinning top	5.2
1903	Boy proves the Gauss-Bonnet theorem for closed surfaces	11.3
1906	Da Rios defines the filament equation	5.3
1923	Thomsen defines Willmore surfaces, then called Konformminimalflächen	13.1
1931	Douglas and Rado independently prove the existence of a minimal surface with prescribed boundary curve	12.4

(continued)

© The Author(s) 2024
U. Pinkall, O. Gross, *Differential Geometry*, Compact Textbooks in Mathematics,
https://doi.org/10.1007/978-3-031-39838-4

Table B.1 (continued)

Year	Milestone	Section
1937	Whitney and Graustein proof their theorem	3.6
1956	Hopf proves that round spheres are the only constant mean curvature surfaces in $\mathbb{R}^3$ of genus zero	12.5
1965	Willmore states his conjecture	13.1
1970	Lawson finds closed minimal surfaces in S^3 with any genus	13.1
1972	Hasimoto shows that the filament equation is a soliton equation	5.3
1984	Wente finds the first constant mean curvature torus	12.5
2012	Marques and Neves prove the Willmore conjecture	13.1

References

1. A.D. Alexandrov, Uniqueness theorems for surfaces in the large, I. Vestnik Leningrad Univ. **11**, 5–17 (1956). English translation: Amer. Math. Soc. Transl. 21 (1962), 341–354
2. G. Arreaga, R. Capovilla, C. Chryssomalakos, J. Guven, Areaconstrained planar elastica. Phys. Rev. E **65**, 031801 (2002)
3. G. Bardini, G.M. Gianella, A historical walk along the idea of curvature, from Newton to Gauss passing from Euler. Int. Math. Forum **11**(6), 259–278 (2016)
4. M. Bergou, M. Wardetzky, S. Robinson, B. Audoly, E. Grinspun, Discrete elastic rods, in *ACM SIGGRAPH 2008 papers, SIGGRAPH'08*, vol. 27 (Association for Computing Machinery, New York, 2008), pp. 63:1–63:12
5. J. Binet, Mémoire sur l'intégration des équations de la courbe élastique à double courbure. C. R. Acad. Sci. **18**, 1115–1119 (1844). Englisch translation: Neo-Classical physics
6. W. Blaschke, *Vorlesungen über Differentialgeometrie und geometrische Grundlagen von Einsteins Relativitätstheorie III*. Grundlehren der mathematischen Wissenschaften, vol. 29 (Springer, Berlin, 1929)
7. A.I. Bobenko, S. Heller, N. Schmitt. Minimal n-noids in hyperbolic and anti-de Sitter 3-space. Proc. R. Soc. A **475**, 2227 (2019)
8. W. Boy, Über die Curvatura integra und die Topologie geschlossener Flächen. Math. Ann. **57**, 151–184 (1903)
9. W.D. Callister, D.G. Rethwisch, *Materials science-and-engineering: an introduction*, 8th edn. (Wiley, Hoboken, 2009)
10. A. Chern, F. Knöppel, F. Pedit, U. Pinkall, Commuting Hamiltonian flows of curves in real space forms (2018). arXiv: 1809. 01394 [math.DG]
11. P. Dombrowski, Differentialgeometrie, in *Ein Jahrhundert Mathematik 1890–1990: Festschrift zum Jubiläum der DMV*, ed. by G. Fischer, F. Hirzebruch, W. Scharlau, W. Törnig (Vieweg+Teubner Verlag, Wiesbaden, 1990), pp. 323–360
12. J. Douglas, Solution of the problem of Plateau. Trans. Am. Math. Soc. **33**, 263–321 (1931)
13. L. Euler, Methodus inveniendi lineas curvas maximi minimive proprietate gaudentes, sive Solutio problematis isoperimetrici latissimo sensu accepti. Apud Marcum-Michaelem Bousquet Socios Genevx (1744). https://doi.org/10.5479/sil.318525.39088000877480
14. V.G.A. Goss, Snap buckling, writhing and loop formation in twisted rods. Ph.D. Thesis. University of London (2003)
15. H. Hasimoto, A soliton on a vortex filament. J. Fluid Mech. **51**(3), 477–485 (1972).
16. M.W. Hirsch. *Differential topology*. Graduate Texts in Mathematics, vol. 33 (Springer, New York, 2012)
17. H. Hopf, *Differential geometry in the large*. Lecture Notes in Mathematics, vol. 1000 (Springer, Berlin, 1983). Seminar Lectures New York University 1946 and Stanford University 1956
18. L. Hsu, R. Kusner, J.M. Sullivan, Minimizing the squared mean curvature integral for surfaces in space forms. Exp. Math. **1**(3), 191–207 (1992)

U. Pinkall, O. Gross, *Differential Geometry*, Compact Textbooks in Mathematics,
https://doi.org/10.1007/978-3-031-39838-4

19. G. Kirchoff, Über das Gleichgewicht und die Bewegung eines unendlich dünnen elastischen Stabes. J. F. D. Reine U. Angew. Math. **56**, 285–313 (1858)
20. D. Kleckner, M.W. Scheeler, W.T.M. Irvine, The life of a vortex knot. Phys. Fluids **26**(9), 091105 (2014)
21. R. Kusner, Conformal geometry and complete minimal surfaces. Bull. Am. Math. Soc. **17**(2), 291–295 (1987)
22. J.L. Lagrange, *Mécanique analytique*, vol. 1 (Chez la Veuve Desaint, Paris, 1788)
23. H.B. Lawson, Complete minimal surfaces in S^3. Ann. Math. **92**(3), 335–374 (1970)
24. T. Levi-Civita, Attrazione newtoniana dei tubi sottili e vortici filiformi. Annali della Scuola Normale Superiore di Pisa - Classe di Scienze **1**(3), 229–250 (1932)
25. R.L. Levien, From spiral to spline: optimal techniques in interactive curve design. Ph.D Thesis. University of California, Berkeley (2009)
26. F.C. Marques, A. Neves, Min-Max theory and the Willmore conjecture (2012). arXiv:1202.6036
27. F.C. Marques, A. Neves, The willmore conjecture. Jahresber. DMV **116**(4), 201–222 (2014)
28. J. Marsden, A. Weinstein, Coadjoint orbits, vortices, and Clebsch variables for incompressible fluids. Physica D **7**(1), 305–323 (1983)
29. A.F. Möbius, Theorie der elementaren Verwandschaft. Berichte über die Verhandlungen der königlich sächsischen Gesellschaft der Wissenschaften zu Leipzig, mathematisch-physische Classe **15**, 1–16 (1862). Reprint: A.F. Möbius, Gesammelte Werke V.2, Leipzig 1886, pp. 433–471
30. I. Newton, *Method of fluxions* (Henry Woodfall, London, 1736)
31. R. Perline, Poisson geometry of the filament equation. J. Nonlinear. Sci. **1**, 71–93 (1991)
32. T. Rado, On plateau's problem. Ann. Math. **31**(3), 457–469 (1930)
33. R.L. Ricca, Rediscovery of Da Rios equations. Nature **352**, 561–562 (1991)
34. R.L. Ricca, The contributions of da rios and levi-civita to asymptotic potential theory and vortex filament dynamics. Fluid Dyna. Res. **18**(5), 245–268 (1996)
35. F. Schmelz, H.C. Seherr-Thoss, E. Aucktor, Theory of constant velocity joints, in *Universal joints and driveshafts: analysis, design, applications* (Springer, Berlin, 1992), pp. 29–56
36. M. Spivak, *A comprehensive introduction to differential geometry*, vol. 2 (Publish or Perish, 1999)
37. D. Swigon, The mathematics of DNA sturcture, mechanics, and dynamics, in *Mathematics of DNA structure, function and interactions*, ed. by C.J. Benham, S. Harvey, W.K. Olson, D.W. Sumners, D. Swigon (Springer, New York, 2009), pp. 293–320
38. G. Thomsen, Über konforme Geometrie, I. Grundlagen der konformen Flächentheorie. Abh. a. d. Math. Sem. d. Hamb. Univ., Bd. **3**, 31–56 (1923)
39. W. Thomson, Vibrations of a columnar vortex. Proc. R. Soc. **10**, 443–456 (1880)
40. E.H. Tjaden, Einfache elastische Kurven. Ph.D. Thesis. Technischen Universität Berlin (1992)
41. L.J. Villegas Vicencio, M.J. Larrañaga Fu, J.R. Lerma Aragón, R. Romo, J. Tapia Mercado, Precession and nutation visualized. Lat. Am. J. Phys. Educ. **6**(1), 179–182 (2012)
42. A.H. Wallace, *Differential topology: first steps*. Dover Books on Mathematics (Dover Publications, Mineola, 2006)
43. H.C. Wente, Counterexample to a conjecture of H. Hopf. Pac. J. Math. **121**(1), 193–243 (1986)
44. H. Weyl, *Die Idee der Riemannschen Fläche*. Mathematische Vorlesungen an der Universität Göttingen (B. G. Teubner, Berlin, 1913)
45. H. Whitney, On regular closed curves in the plane. Compos. Math. **4**, 276–284 (1937)
46. T.J. Willmore, Note on embedded surfaces. An. Şti. Univ. "AL. I. Cuza" Iaşi Secţ. I a Mat. **11B**, 493–496 (1965)
47. T.J. Willmore, *Riemannian geometry*. Oxford Science Publications (Clarendon Press, Oxford, 1996)

Index

© The Author(s) 2024
U. Pinkall, O. Gross, *Differential Geometry*, Compact Textbooks in Mathematics,
https://doi.org/10.1007/978-3-031-39838-4